G000242243

WILD
CITY

WILD CITY

Encounters with Urban Wildlife

Florence Wilkinson

First published in Great Britain in 2022 by Orion Spring,
an imprint of The Orion Publishing Group Ltd
Carmelite House, 50 Victoria Embankment
London EC4Y 0DZ

An Hachette UK Company

1 3 5 7 9 10 8 6 4 2

A CIP catalogue record for this book is
available from the British Library.

ISBN (Hardback) 978 1 3987 0185 4
ISBN (eBook) 978 1 3987 0187 8
ISBN (Audio) 978 1 3987 0188 5

Printed and bound in Great Britain by Clays Ltd, Elcograf S.p.A.

MIX
Paper from
responsible sources
FSC® C104740

www.orionbooks.co.uk

In memory of my mother, Sara Wilkinson,
who filled my life with love, imagination and animals.

What is the city but the people?

William Shakespeare, *Coriolanus, Act 3, Scene 1*

Contents

Foreword

A damselfly whirs on intricately laced wings, dipping its slender blue abdomen in and out of the water. A coot forages amongst the pondweed, while its scruffy-looking chick balances on a nest of twigs, mouth gaping as it waits impatiently for its next meal. The water is almost perfectly still, save for the odd bubble when a fish pops up to the surface to snatch at an insect. A weeping willow solemnly bows its head and a cabbage white butterfly rests on a nettle.

Overhead, gulls cry and a swift, living up to its name, darts and wheels through the virtually cloudless sky. The sun beams down on a jet-black cormorant standing on the bank, with wings theatrically outstretched as if about to perform a magic trick. At the water's edge a gang of sparrows chatter querulously, flitting in and out of the undergrowth.

The cormorant fans its wings, feathers not yet fully dry. It turns its head and fixes me with a defiant gaze. We remain locked in this staring contest for several moments, until we're interrupted by a loud clattering. I become conscious of a sharp ringing in my ears and an almighty creature with armoured head, arched back and legs spinning comes hurtling towards me. 'EXCUSE ME', it hollers, seemingly put out. I step back just in time to avoid being mowed down as the Brompton bicycle whooshes past.

The cormorant remains unmoved, but for me the spell has been broken. A woman jogs past, ponytail swishing. A man in a suit marches purposefully on towards his next meeting, while a couple push a buggy in the other direction. Horns beep, trains rumble, motorbikes growl, narrowboats chug – this is rush hour on the towpath of the Regent's Canal, about halfway between London's King's Cross and Camden Town, and five minutes from my home. To my left lies St Pancras International Station, where trains thunder to and from the Continent, to my right the famous Camden Market, which draws in thousands of tourists every day.

If you ask most people where you should go to get closer to nature, they will likely reel off a list of remote or exotic locations – dense woodlands, rushing rivers, open seas; maybe the Serengeti, even, or the Arctic tundra. We use the phrase 'the natural world' as if it exists in a bubble, divorced from our everyday lives. The implication is that nature only belongs in places that are 'natural'. Cities are 'man-made', and therefore inherently unnatural.

Such arguments ignore the fact that we humans are animals ourselves, and in any case nature does not respect these imagined boundaries between the 'natural' and the 'man-made'. Where we see a tower block, a peregrine sees a cliff. What looks to us like an old canal tunnel could easily resemble a cave to a bat. For bees and other pollinators, a hanging basket is a tiny wildflower meadow. Even the famous Trafalgar Square fountains have served as ponds for a passing mallard or two.

Until 2007, more people on our planet lived in rural areas than urban, but in that year, for the first time in human history, the balance tipped the other way. The UN predicts that by 2030 one-tenth of our planet will have been urbanised. By 2050, the world's population is estimated to reach 9.8 billion, with nearly 70 per cent of these people projected to live in urban neighbourhoods. In fact, city-living has long been our preferred lifestyle across Europe and North America – more than eight in ten Brits are urban dwellers, according to government statistics.

Meanwhile, large swathes of the countryside have become inhospitable to our native wildlife due to industrialised farming, land management practices and pollution. These inescapable realities are forcing nature to find new ways of surviving (and even thriving) in an urban environment.

But not everyone is willing to welcome nature in all its guises into our towns and cities. While most people are happy to see a robin on their bird feeder, far fewer will roll out the red carpet for the pigeons in their local park. Feral pigeons derive from domestic pigeons, which were bred from wild rock doves. They have adapted especially well to city life, taking up residence in spaces that were never intended for their use. And now businesses install mean-looking spikes to deter them, people allow their children and dogs to chase them, councils put up nets in which they often become hopelessly entangled.

Other less-than-welcome urban visitors include gulls, which have taken to snatching food clean out of our hands, and grey squirrels, which come under fire for displacing our native reds. Anyone who has recently visited a park or green space of some description in our capital city has probably also encountered the thriving population of ring-necked parakeets. Easily identified by their piercing shriek or lurid green plumage, there have been calls for a mass cull by those who argue that they do not belong here.

Living in ever-closer proximity creates new tensions in our relationship with nature. Anything that doesn't conform to our need for cleanliness and order – or transgresses the artificial limits we place upon it – is sanitised out of existence. Parks are pruned and manicured. Gardens are decked or paved over. Yet at the same time we increasingly feel a longing to connect with our wilder roots and rebuild a little wilderness in our own back garden, if we're lucky enough to have one. Our idealised version of nature doesn't always match up with the messy reality. But when our towns and cities have already taken over so much of the world's liveable space, is it right that we should become the self-appointed arbiters of what constitutes 'wildlife' and what becomes a 'weed' or a 'pest'?

This proximity also offers opportunities to engage with nature in new ways. Growing up in semi-rural Essex, I'd never seen a kingfisher up close, or come into contact with a cormorant, yet I've gained both these experiences since living by the canal in central London.

Urban geographers and sociologists have spilt plenty of ink debating what we mean by a city. In some parts of the world, cities are defined by the density of their population. Here in the UK, city status is mostly administrative; our seventy cities have been granted their status by the monarch and can claim certain state benefits as a result. I've visited a number of them as part of my research, but I've also taken the liberty of including examples from towns too, my interest being in wildlife rather than technicalities.

This book is a journey into why we should embrace, celebrate and support nature in our cities. I'm not a biologist, botanist, ecologist or entomologist (although I could now introduce you to plenty of each). I'm just someone who loves both city life and wildlife in equal measures. And as a writer and journalist, I've used what comes naturally to me – researching, asking lots of questions and listening – to explore how wildlife is changing and adapting to an increasingly urban environment, as well as the actions we can take both individually and collectively to help it thrive. Like planting bee-friendly flowers, creating a small garden pond, or generally leaving our green spaces to grow uninhibited, resisting the urge to strim and mow.

Nature is not a moral force, but as conscious beings we can acknowledge that we're not the only species to which this planet belongs, and we have the power to undo at least some of the damage we've already wrought upon it. Promoting urban wildlife benefits our own species, too – for better or worse, we are all affected by life in the city. We've long known the negative impact a lack of access to adequate green spaces can have on our mental health and that of our children. Engaging with urban nature can provide a vital point of reconnection that so many of us crave. And it comes back full-circle; without this connection, how else do we persuade future generations to protect the wildlife on their doorstep and beyond?

I began writing *Wild City* before all of our lives changed in ways we could never have imagined, as the coronavirus pandemic swept through this country and across the globe. I continued my research through the multiple lockdowns we endured, adjusting to the reality of self-isolation, or the 'new normal', as it euphemistically became known, our worlds shrinking down to the hyper-local. As part of this process we became more aware of our surroundings and the preciousness of our urban green spaces. Friends got in touch to tell me about the birds visiting their gardens or balconies, and we all took solace in the renewal of spring in 2020 like never before. An initial reduction in traffic and noise pollution gave our ears the chance to re-tune to the sounds of nature in the heart of the city.

As our worlds grew smaller – restricted to our homes and perhaps a daily walk around the block – the wild creatures living nearby became more notable, meaningful, essential. Never had I more greatly appreciated the nest of bumblebees travelling back and forth on their quest for pollen through a hole at the bottom of our fence. Blue tits, great tits, robins, blackbirds, thrushes, sparrows, jays – all of a sudden I was able to develop a relationship with the birds in our garden, attending to their preferences, likes and dislikes and growing accustomed to their behaviours. After a rather unpleasant personal encounter with COVID-19 over Christmas, the first thing my partner Ben and I did once our allotted ten days of house arrest were up was to go for a long walk on Hampstead Heath, where we took in big gulps of the icy air, savouring the delicious crunch of our boots stomping through the frozen grass.

As a final note, I don't want anyone to think I am suggesting that with a few small tweaks here and there nature will be just fine, while urbanisation continues unabated. Or that we don't need to concern ourselves with the impact of habitat loss and the tree-felling, exhaust-spewing, air-polluting consequences of urban living. Far from it. Like anyone who cares about biodiversity and the environment, I intensely believe that we must do everything in our power to preserve those remaining pristine habitats that still exist on this planet. Like others, I shed a tear when I read that the US will be using the beautiful island of San Cristóbel in the Galápagos, which I was once lucky enough to visit, as an airfield for its military planes. I felt the same pain when images emerged of a polar bear wandering listlessly through the streets of a Siberian city, forced hundreds of miles away from its home through hunger and desperation.

But when we talk about habitat loss, we should also look at habitat gain, and here our cities offer great potential. Though rooftops, canals and remnants of industry might be of human construction, they can provide the conditions for many species to survive and flourish.

In 2015 I launched a birdsong recognition app called Warblr with my co-founder Dr Dan Stowell. People soon started telling us that after downloading our app they kept hearing birds. In their back gardens. At the train station. While walking through the streets of London, Bristol, Birmingham and Edinburgh. Most of these people hadn't expected to find wildlife in their city, so they had somehow tuned its frequency out of their everyday lives. But in truth it was there all along.

CHAPTER I

Welcome to the Urban Jungle

From: ███████████████████████
Sent: 18 December 2014 10:26
To: ████████
Subject: Foxhunting

Dear Neighbours,
Coming home late last night, I was astonished to find
a resident of ████████ Road deliberately feeding a fox
on the pavement. As several emails have shown, these
verminous animals are coming into our houses, stealing
food, and infecting our own domestic animals with
mange and other diseases. I would like to enter a plea to
anyone else who may be encouraging their local existence, to think twice. It is tantamount to feeding rats.

In fact, for some time I have been wondering about
instituting a hunt, for those of us equipped with
bicycles and dogs, perhaps operating by night. A Boxing
Day Meet might be a good Christmas tradition to
inaugurate this year.

Season's Greetings
████████████████

*Email from a north Londoner to their local residents'
association*

* * *

ESCAPE TO THE CITY

I'm standing on an unremarkable street, just five minutes from the estate where I live. To my right is the towering eighteen-storey council block that I can see from my back garden. Facing me is a slightly tired-looking period conversion – the kind you find across much of London. Two 360-degree CCTV cameras blink up from the basement flat. To the side of the building, in between a jumble of plants, compost sacks and garden waste bags, is a mesh crate with a black metal frame and a mechanical trap door that's slightly lifted. On top of the crate is another CCTV camera, encased in Perspex to protect it from the elements.

'So this is where Derek lives. He's probably watching us on his cameras now,' Karen states matter-of-factly. 'Watch where you step,' she cautions, gesturing at the pavement; 'that's probably fox poo there.'

I got in touch with Karen, the founder of my local fox rescue, some time ago, hoping to get her thoughts on why we've seen such an increase in urban foxes over the past few decades. She has been hard to pin down, despite living just around the corner – caring for Camden's wildlife is the equivalent of at least two full-time jobs, it seems. In addition to foxes she rescues pigeons, feral cats and pretty much anything else she comes across. After a lot of back and forth, Karen finally suggested I could join her on one of her regular walks around the neighbourhood to see how its non-human inhabitants are faring. I tell her I wasn't expecting her to be so well groomed, as someone who spends their time literally picking unfortunate creatures up off the pavement. She laughs.

The contraption we're looking at, Karen explains, is of Derek's own making. It's designed to administer medication for the treatment of mange – the second-biggest killer of urban foxes, after traffic. Derek is pretty much housebound, but he has found an ingenious way to help foxes with the infectious skin disease, which is caused by parasitic mites and is more common in urban animals that live in much closer quarters than their rural cousins. Treating foxes is tricky; the best method is to conceal the medication in some food, but since these are wild animals there's no guarantee that the right fox (or even the right species) will consume it. Derek's invention solves this conundrum. He uses his cameras to watch what's going on in real time and is then able to remotely open or shut the trap door, so that the right fox gets access to the right meds. He's already saved one fox's life using this method, and now he and Karen are looking at using his invention as a prototype which could potentially be rolled out in other areas too.

Karen started rescuing urban foxes a few years ago, when a female with a broken leg turned up in her garden. She soon found herself the first responder for every hapless creature that presented itself thereafter: 'As soon as the neighbourhood got wind, that was it!' she explains, describing how she took on the mantle of north London's foremost fox rescuer. Karen grew up outside the city, and yet her encounters with foxes only began when she moved to London in her twenties. This very much tallies with my own experience. Growing up in semi-rural north Essex, I would on rare occasions spot a shadowy vulpine silhouette through a car window, ears pricked and nose raised skywards. But it wasn't until I moved to London in my early twenties that I met with a red fox up close. After that first encounter, foxes would appear in the corner of my eye as I made my way home after a night out, or early in the morning when I was putting out the bins.

When I moved into a flatshare in Kentish Town, a fox family moved in too. Into the garden, at least. Hearing the males' strange human-like screams during mating season, we initially considered calling the police. 'It's mating season at the moment,' Karen observes, shaking her head; 'their hormones are running wild. We find many more dead on the road at this time of year.'

Living in such close proximity to these wild canids can result in some interesting experiences. One fine summer's day a few years back, my former flatmate Nick went to take his morning shower. Pulling back the shower curtain, he found himself confronted by a young fox hiding in the bathtub. The night before, someone had left the garden door open to let in some air. The bushy-tailed interloper must have made its way up the wrought-iron staircase – which we struggled to negotiate ourselves after a few drinks – into the living room, up a further flight of stairs and through the bathroom door. After a call to the RSPCA for advice, Nick used an old mattress as a makeshift ramp, allowing the fox to climb out and make a hasty exit.

If housebreaking wasn't daring enough, in February 2020 a fox made it through one of the nation's most advanced security systems and into the Houses of Parliament. The creature reached the fourth floor of Portcullis House before police, reportedly wearing large gloves, were finally able to capture it in a box and set it free nearby.

The rise of the urban fox is a relatively recent phenomenon and helps shed light on how some animals are carving out a niche for themselves in towns and cities. The occasional fox was recorded on London's Hampstead Heath from 1912 onwards, but regular sightings only began in UK towns and cities in the 1930s. During the First and Second World Wars, and the period in between, urban foxes really began to flourish, with housing developments, the rise of suburbia and increasing numbers of large gardens all contributing to their success. Cities offer valuable shelter, including plenty of safe places to reproduce. While country-dwelling foxes expend considerable time and precious energy digging into the ground or under tree roots to create their dens, their urban counterparts have quite literally got it made. Sheds, basements, sub-floors and decking only need small modifications before they can be converted into a snug little home, ideal for raising the cubs.

By the 1940s there were enough foxes in the capital for the government to deem them a problem. Foxes became victims of their own success, and hunters were hired as part of an eradication programme that ran for thirty years, but with little discernible effect. Today there are an estimated 150,000 urban foxes in England. The highest concentration can be found in Bournemouth – twenty-three foxes per square kilometre – followed by London, Bristol and then Newcastle.

Nor is this urban success restricted to British foxes – foxes have taken up residence in cities worldwide, in both their native ranges and in non-native areas where they've been introduced. Populations of urban foxes can now be found across Europe, in Australia, Japan and North America – wherever they find an opportunity they seem to take it up, making them the world's most successful wild carnivore.

The field of urban ecology traditionally places species within three categories: urban avoiders, urban adaptors and urban exploiters. Urban avoiders cannot adapt to an urban environment, often having complex and specific needs for their lifecycles. They usually suffer the most from growing urbanisation, and are at the greatest risk of extinction as their habitats are overwhelmed, ornamentalised, fragmented and destroyed – species such as the rare sand lizard, which can only reside in sandy heathlands or dunes, the brown hare or the turtle dove. Urban adapters do well in moderately developed areas and have learnt to live with us humans; they are opportunists that will gladly populate our gardens. These are creatures such as robins, great tits and badgers which, on the whole, are met with a warm reception from humans. Urban exploiters are those hardy omnivores that can thrive in the most concrete of jungles, our cities' permanent residents – pigeons, sparrows, mice and foxes. Not all so warmly welcomed by us.

Living in increasingly close quarters with humans also means that many species of urban wildlife have become accustomed to our presence and learnt that most city dwellers are unlikely to prove a threat. Studies show that urban blackbirds, for instance, are less afraid of humans than their rural counterparts (and may even be evolving towards a separate species). They'll happily raid bins in the local park when it's filled with people, or take food from garden bird feeders. Both blackbirds and foxes are adaptable animals: they are able problem solvers with what scientists call 'behavioural plasticity' – which is key to their urban success.

And while the average lifespan of an urban fox is just eighteen months – sadly, large numbers get killed on the road – city living clearly has enough appeal to make it worth their while. Karen tells me that, contrary to popular belief, she never sees a starving fox, so great is the abundance of food available to them. Foxes are smaller animals than many people realise – long-legged and with sizeable tails, they give the impression of being greater in stature than they really are. So when people see a fox close up they often mistakenly think it's underfed. In fact, the average female weighs just 5–6 kilos, and the average male around 6–7 – not much bigger than my Pomeranian dog (whom, as it happens, we named Fox, due to her passing resemblance to these fellow canids). 'They do get skinny and they lose a lot of weight if they've got mange, because malnourishment is part of the problem,' Karen explains, 'but that's not for want of food. Foxes are self-regulating – there wouldn't be so many of them if they weren't thriving – I mean, there's rats, birds, mice everywhere . . .'

She pauses, glancing down at the ground and screwing up her face; as if to prove her point, there's a half-eaten mouse carcass on the pavement. 'And then they eat worms, protein, berries, fruit,' she adds, stepping tentatively around the remains of the mouse, 'and all the rubbish we leave around. People feed them too.'

Like most urban exploiters, as generalists – unfussy and unspecialised eaters – foxes are thriving on the all-you-can-eat buffet that cities provide, whether they fancy a spot of fried chicken, a strawberry Frappuccino or a mouse kebab.

TRAVEL COMPANIONS

There's a fierce chill in the air. The ground itself is frozen and crunchy underfoot, veiled by a thin layer of grasses, mosses and lichens, which form a carpet of ochres, oranges and reds. Only the odd shrub at ankle or knee height serves to break up the treeless landscape.

These days best known for its football club, 440,000 years ago Tottenham, in the north London borough of Haringey, was a very different place. Not far from the Tottenham Hotspur grounds lay the toe of a vast sheet of ice of up to 2 kilometres thick at its greatest extent. The ice sheets of this period held so much frozen water that global sea levels were over 100 metres lower than they are today. For much of this time, the North Sea ceased to exist at all, allowing animals, including early people, to wander all the way across from Africa virtually uninterrupted.

When the first humans arrived here, following a long and gruelling migration, they were short and stocky – well-adapted to cope with extreme environments – their brows more pronounced than our own and their chins rather less so. Though we're related, *Homo heidelbergensis* was a different species from our own. They hunted across the tundra, using flint-headed spears and hand-axes made from nodules of rock. The creatures they sought were huge and lumbering, liable to charge and stampede: woolly mammoths with oversized tusks and woolly rhinos with reddish-brown coats; musk ox, bison and bears, as well as smaller and less risky quarry such as deer.

Fast-forward another 320,000 years, to 120,000 years ago. The people are gone from what will one day become London – both *Homo heidelbergensis* and the later arrivals, the Neanderthals. Prehistorians aren't sure why humans were absent during this period, but there's no evidence of their presence in the fossil record. They're missing out on an age of exceptional warmth, with average temperatures reaching 2–3 degrees higher than today. Across the savanna-type landscape roam straight-tusked elephants, along with the woolly rhinos and bison. In what is now Trafalgar Square the lions are real, not statues, as are the hippopotamuses, which wallow contentedly

11

in the space where the two fountains now stand. We know this because remains – both lion and hippo – have been uncovered in the area, a discovery which turned the centre of our capital city into the most important prehistoric site in Britain for this period.

Leaping onwards by 114,000 years, which takes us to 6,000 years ago, things have changed significantly in pre-historic London. The savanna is gone, the tundra is gone and the future city, like the rest of the country, is cloaked in broadleaf deciduous woodland. The human settlers – for they are now finally starting to settle, rather than merely roaming through – are venturing into farming, with herds of domestic animals, although their agriculture is still supplemented by hunting. Brave gangs return home after an expedition with wild boar, red and roe deer and aurochs – wild cattle with sizeable horns, nearly 2 metres high at the shoulders, the long-lost ancestors of domestic cattle.

Our landscape and its accompanying flora and fauna has always been in a state of flux as temperatures have fallen and risen again, but it was during this period that we humans slowly began to make our mark on the land, felling trees and hunting to extinction most of our remaining megafauna (those giant animals that no longer grace these shores and, where they do still exist, are under constant threat across their remaining global ranges). Over time, our ancestors learnt how to cultivate crops and successfully rear livestock, until they could produce enough from one portion of land to finally abandon the hunter-gatherer lifestyle. They supplemented what they raised and grew by foraging locally for items such as nuts, berries, fungi and medicinal herbs. They formed small settlements and set about clearing more and more of the land for agriculture and an increasing population.

It wasn't until the Romans arrived in Britain that we could begin to describe any of these settlements on our islands as urban. But in Mesopotamia, which comprised present-day Iraq and parts of Iran, Turkey, Syria and Kuwait, the increase in productivity brought about by the Neolithic revolution led to more people living together in higher densities. This population density resulted in new benefits for humans, such as the sharing of ideas and resources, communal care for children and the elderly, the development of markets, reduced transportation costs and access to amenities like running water and sewage disposal. These were the first cities.

The rise of the human city affected other species in different ways. Habitat loss drove many into extinction, but there were others that found our company beneficial: those that actively sought out an association with humans and the habitats we create. Some creatures took it one step further, spending their whole lives within the city limits and even developing a commensal relationship with people, becoming fully reliant on us for morsels of leftover food, and on our buildings for shelter. (The word commensal comes from the Medieval Latin *commensalis*, literally meaning to 'eat at the same table'.) On the whole, these commensal species, such as rats and mice, caused neither direct harm nor benefit to us. They tended to be smaller,

hardy, adept at adapting, with high reproductive capa-
bilities, and for better or worse our homes became their
homes. These fellow travellers have gone on to colonise
each and every corner of the globe with us, playing their
own part in shaping the ecology of our urban centres.

On 5 August 2019 I was sitting in a café, drinking a hot
chocolate and watching one such commensal creature – a
house sparrow, as it carefully picked at some crumbs near
my feet.

My local congregation of sparrows has taken to
spending all day long on our bird feeder, elevating our
commensal relationship to new heights in which I have
become their willing servant, changing their water twice
daily and furnishing them with vast quantities of suet balls.
Sparrows aren't nearly as common in the UK as they once
were – ecologists aren't entirely sure why, but over the past
few decades their numbers have been declining. In the
capital this began in the late 1920s, but the rate of decline
increased in the 1990s to the point where they disappeared
in many parts. Only in the past ten years have we seen a
partial reversal of this decline. London's Cockney Sparrows
– both the human East Enders and the birds after which
they were nicknamed – are much rarer than they were in
my childhood, so I'm keen to support them.

But that's not why this particular encounter in the café
was surprising, nor why I kept a precise note of the date.
It was surprising because of where the café was situated:
far from London, the UK and from Europe for that matter,
2,182 miles from the nearest continental point, on Easter

Island, where house sparrows have been introduced by humans. Watching this little brown bird far from home with the crashing waves of the South Pacific as its backdrop, I felt a sense of belonging – something we lose when we separate ourselves from nature – and a connection with our shared histories, which are closely interwoven.

As far as we know, house sparrows have lived with humans ever since we began to live in houses, or structures of some sort, feeding on whatever we leave behind. Sparrows, or rather their close ancestors, first appear in the fossil records around 10,000 to 20,000 years ago in what is now modern-day Israel – not far from where those first cities were founded in Mesopotamia. They made their way to Europe with humans and the spread of agriculture, and from there they've followed us ever since, even to Easter Island.

Ironically, it was another commensal creature that may have caused the destruction of Easter Island's precious resources, and much of the human suffering that followed. It's still widely believed that the Rapanui people were responsible for the fall of their own civilisation, as they chopped down the island's lush green palm trees to transport their famous statues, the vast stone heads that we associate with the island. A lesser-known fact is that when the settlers first arrived at this uninhabited island, between 600 and 900 AD, they brought with them Pacific rats in their canoes. Whether this was unintentional or planned (perhaps as an emergency protein source for the journey), the arrival of rats had an undeniable impact on the ecology of the island. While the human population grew to around 12,000 at its peak the rats, with no real predators, are thought to have reached a number of several million. They developed a taste for the seeds of the unique Easter Island palms, which most likely prevented the trees from reproducing. By the time the islanders began removing ever-greater swathes of trees to transport their statues, the real damage to the island's delicate ecosystem may well have already been done.

Another species that evolved alongside human settlements is the ever-ubiquitous house mouse. Thought to have originated in parts of India, house mice spread to the eastern Mediterranean about 15,000 years ago and have followed humans ever since. The tiny rodents didn't arrive in the rest of Europe until around 1000 BC; the lag is thought to be due to their need for urban density – human settlements must be above a certain size for them to thrive. In the early sixteenth century, house mice arrived in the Americas on the ships of Spanish explorers and Conquistadors, and 100 years later the British, along with French merchants and traders, transported them to North America.

As their name denotes, house mice have become reliant on us for their very existence. Their needs are modest – to stay warm, dry and to eat just three grams of food a day (contrary to popular belief, their preference is for cereals and grains, not cheese). And as masters of adaptation they will burrow, gnaw, hide and climb in order to fulfil these needs, nesting in wall voids, cellars, storage spaces and lofts. There's a population today that live out their whole lives on the London Underground; a friend's father, who worked for the travel network, reported witnessing passengers falling on to the tracks while watching the mice instead of minding the gap.

Though the lions and hippos and aurochs are long gone from British cities, through the movements of our much smaller, more adaptable commensals it's possible to trace our own migrations. Taking advantage of the urban environment, they have adapted with us, and their original niche may have now disappeared. Evolution, of course, is a fluid process, so many more plants and animals may come to rely on our cities in the future too. The idea that we've ever been somehow sheltered or separate from other wild creatures is an urban myth. They have been living alongside us all along.

AN URBAN NATIONAL PARK

It's a drizzly day in January – not so wet as to justify an umbrella, but persistent enough to cause a constant dampness on the surface of the skin. Geographer Dan Raven-Ellison has asked me to meet him by the Narrowboat Pub on the Regent's Canal, in between north London's Islington and King's Cross.

Dan and I met a few years ago when he asked me to speak about my birdsong recognition app, Warblr, at an event he was organising. He has been looking at cities differently ever since he embarked on a project to visit every single one of the UK's national parks over a six-month period. As is so often the case, being away made him think about home, and he began to reflect on the relative value of the city where he lives as a space for nature. 'It was during that journey,' he tells me, 'that I began to wonder why it is, when you look at the family of national parks around the world, they represent every major type of internationally recognised landscape apart from the fastest-growing landscape in the world – urban areas.'

It's very true – when I was a child I had a Dorling Kindersley encyclopaedia of nature in the form of a CD-ROM, which I would insert into my dad's old Apple Macintosh computer. It was organised around the five great ecosystems of our planet – aquatic, desert, forest, grassland and tundra. Cities didn't even get a look-in.

Dan was so struck by the idea that urban environments could and should be valued as part of the natural world that he spent the next few years campaigning to get London recognised as the world's first 'National Park City' (a phrase which some people might consider an oxymoron). In July 2019 he succeeded, when London Mayor Sadiq Khan held a summit at City Hall to make the announcement, which was followed by a National Park City Festival. The campaign has now spawned a foundation, encouraging other cities across the globe to follow suit.

Wearing a sensible rain jacket and comfy shoes, Dan is much better prepared for our wet walk than I am. Since we last met he's gained an impressive beard, which is now collecting tiny droplets of rain that glisten as he speaks. As is often the case, I'm less suitably dressed for inclement weather, in heeled boots with no grip and a bright pink coat that isn't nearly warm enough, which I'm supposed to be reserving for 'smart'. My mother used to despair at my wardrobe choices – when I went travelling after university I packed high heels over walking boots and two blazers instead of a waterproof.

We follow the canal for just a few minutes before reaching the spot where it moves underground (only boats can pass through the tunnel), so we climb the stairs back up to street level, preparing to rejoin the towpath when it re-emerges. Despite my living nearby and Dan having previously rented a flat in the area, we quickly find ourselves lost in the surrounding streets, the canal beneath our feet, wandering in entirely the wrong direction. This detour sparks a conversation about how we might make our cities greener. 'Most city planners only think in terms of horizontal planes,' Dan observes, 'but just look at the ivy growing up the front of those houses – what a difference it would make if we planted vertically more often – think of the food and shelter it could provide for birds and other animals.'

After finally returning to the canalside we make it as far as King's Cross – so not very far at all – but at least we now have our bearings. This area has changed unrecognisably even in the seven years that I've been living here. Disused coal sheds have gained a new lease of life as swanky boutiques and bars, designed to entice well-heeled Eurostar passengers and tech-company workers; both Facebook and Google are moving en masse to purpose-built offices close to the station. The towering old gasholders have been transformed, with swish new apartment blocks erected inside the vast cylindrical frames (starting at £800,000 for a one-bed). Even the small-but-perfectly-formed Camley Street Natural Park, run by the London Wildlife Trust, has undergone a makeover with a stylish new visitor centre.

Dan and I gaze up at one of the new blocks of flats that seem to have sprung out of nowhere, such is the speed of development here. 'You could compare the side of this building to a windswept cliff face in a glacial national park,' Dan muses; 'both have their own attributes, but actually the "cliff face" of this urban canyon in London may be richer in pickings for a peregrine falcon that's looking for pigeons.' I'm reminded of a recent, unappealing incident in which I witnessed a herring gull dashing a pigeon off the side of this very building and onto the road beneath, before feasting on its still-warm body. 'So just taking rock for rock, surface for surface,' Dan continues, 'I would argue that the urban environment always has value, no matter how gnarly and aesthetically unpleasing an individual may find it.'

Since we're now nearing my house, Dan heads off towards Paddington, with quite some distance left to go before it turns dark. Meanwhile I return home, continuing to muse on urban cliffs, vertical planting and the different dimensions of our National Park City.

The following day I head to London's Natural History Museum to meet with Dr John Tweddle, an ecologist and head of the museum's Angela Marmont Centre for UK Biodiversity. John is a kind-looking man with a neatly groomed beard, bright blue eyes and the slightly cautious manner of someone who thinks carefully before committing to anything. 'In London alone, over 15,000 species of multi-celled life have been recorded,' he informs me; 'that's 20 per cent of all species found across the UK as a whole, and more are being spotted every day.'

Peering through the glass at the Natural History Museum's carefully stuffed and mounted specimens is, ironically, the closest many urban-dwelling children get to experiencing nature in the city. But John, who leads the work here on urban wildlife and citizen science, is doing his best to increase engagement not just within, but beyond the museum's four walls.

John starts our tour by taking me behind the scenes into the cocoon-shaped Darwin Centre, where the collections that are not on public display are stored, to show me some urban-dwelling plant and insect specimens. The room is carefully climate-controlled, maintaining a low temperature and a very low humidity – I can feel the dryness at the back of my throat instantly. You can get a headache from spending too long in here, John warns. These conditions are necessary to safely preserve the nation's natural history collections from pest species that might otherwise chomp their way through the archives. It's impossible to prevent this altogether, but keeping the room cold and dry at least reduces the chances of any such creatures reproducing here. (For the same reason, John asked me to take off my coat before entering in case I'm harbouring any stowaways.)

The collections are stored in rows of tall metal vaults, which give added protection in case of a fire or flood. Rather than looking for urban species in particular, John suggests we start by taking things out at random to see what we find. He turns a metal wheel, opening one of the vaults containing British flowering plants, and pulls out a large folio. Attached to each page are plant specimens that have been dried and pressed, much like I used to press flowers as a child.

Turning to the first page, we're confronted by a rather sad, weedy specimen of a plant known as yellow-rattle. It might not be much to look at when it's dried and pressed, but yellow-rattle plays an important part in conservation. Named after the rattling sound its tiny seeds make as their brown pods shake in the summer breeze, yellow-rattle is a partial parasite of grasses, which is what makes it special. Through preventing grasses from taking over by attacking them at root level, it allows wildflower meadows to flourish – like those being replanted throughout London's Royal Parks and urban rewilding spots across the country. It can even be found in the Natural History Museum's own wildlife garden.

John wants to show me something a bit fancier next, so he suggests we move over to the orchid section. I find myself wondering how he's going to relate this to urban wildlife – to my mind orchids are strange and exotic plants in alien shapes and colours, prized by obsessive collectors with large greenhouses. I've been given orchids in the past, and if I'm honest I'm not fond of them – they're temperamental things and too gaudy for my taste. The folio John pulls out contains bee orchids. Although the specimens themselves don't do this plant much justice – orchids lose their colour when they're dried – bee orchids are truly impressive plants that often appear on nature programmes, a perfect example of one species masquerading as another. Their flowers, which branch from a single upright stem, comprise three pink petals, with a 'bee' hovering in the centre as if collecting pollen.

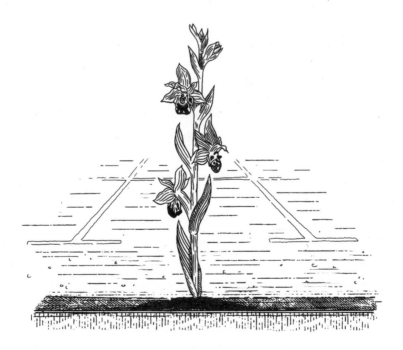

I had always assumed that bee orchids, or any wild orchid species for that matter, must be exceptionally rare. Not so, says John. Like yellow-rattle, bee orchids thrive in grasslands, and as a Mediterranean species they are actually benefiting from our climate becoming warmer and drier, especially in cities. 'They're pretty tough little things,' John tells me; 'I've seen them in Tesco car parks. In fact, we get them across London – we have them here in the wildlife garden too, but we planted them ourselves, so that's cheating!'

We move on to the insect section, where John pulls out a drawer of moths and caterpillars from another row of vaults. 'The colours on all of these caterpillars are real – they're not painted or anything,' he explains; 'they've just been dried very carefully – blow-dried, if you like. A caterpillar is essentially a tube with a head at one end and, well, the opposite at the other, so if you put warm air through it you can dry it from the inside out.' In my head I imagine a miniature hair dryer inflating the caterpillar like a wind sock.

These specimens are of hawk moths, including the spectacular elephant hawk moth, which can be spotted after dark in urban gardens towards the end of the summer. Named after its caterpillar, which looks uncannily like an elephant's trunk and can grow up to 8.5 centimetres long, the adult moth has an olive coat, daubed with psychedelic pink lines and dashes. Also part of this collection is the delicate hummingbird hawk moth. Often subject to a case of mistaken identity as actual hummingbirds (even though hummingbirds don't live in Britain), its wings beat so fast that they produce an audible hum as it darts from flower to flower. Hummingbird hawk moths are migratory, so you may see considerably more individuals from one year to another. They feed in the daytime, making them particularly easy to identify; I've spotted them across the city in parks and pub gardens.

Both John and I are starting to get a bit chilly in the archives, so he suggests we grab our coats and head out to the museum's wildlife garden, a charming patchwork of lowland habitats lovingly recreated in miniature, here in one of the most polluted corners of London. It's a dull winter's day, but amidst the constant roar of traffic I can hear robins and blackbirds singing and the trickling of a tiny ripple pool where honeybees stop to drink in the summer.

In the twenty-five years since the garden was first established, the museum's resident naturalists have recorded over 3,300 species of wildlife here. It provides a valuable example of the mosaic of often isolated habitats that form our urban ecosystem; from remnants of ancient woodland to ponds, gardens and the surrounding roads, pavements and buildings.

It is a common misconception that such diversity of life is only found in the countryside. Conversely, much of our British 'countryside' is dominated by farmland, which can be even less welcoming to wildlife than the most urban of environments; especially following the widespread removal of trees and hedgerows. Farmland is often restricted to just one type of crop, with the use of pesticides and weed killers removing any remaining signs of life. A standard, comparably sized piece of arable farmland or pasture is likely to be considerably less rich in terms of plants and insects, and the animals coming to feed on them, than the Natural History Museum's small one-acre inner-city garden.

The notion that there are any remaining portions of the British Isles that haven't been altered by humans in some shape or form is another misconception, often driven by a nostalgic pastoral idealism which holds that only in the 'wilderness' will we find true nature. Implicit in this idea is that the urban landscape is somehow of lesser value, so we may as well do with it as we please. The truth is that humans have changed the entire landscape, urban or rural, beyond recognition, our fingerprint extending far and wide. You'd be hard pressed to find a single scrap of British land that isn't controlled, owned or managed in one way or another by humans.

As I thank John for his time and head home, I consider some of the other creatures on this planet – ants, termites, beavers and corals – that significantly alter and modify the environment around them. And yet no other species does so with such speed and on such a scale as we humans. Chemists and geologists have now designated a new epoch – the Anthropocene – to describe the vast and unmatchable impact our activities are having on the planet. Some argue that this period started in the late eighteenth century with the invention of the steam engine and the rise of industrialisation. Others contend that it truly began from 1945 onwards, with the seismic growth of the human population and economic expansion, as well as the dropping of the first atomic bomb, when humankind demonstrated its ability to destroy much of life on earth, if we chose to do so. But either way, most agree that the rising levels of carbon dioxide triggered by our own actions will one day show up in the fossil records.

What sets our urban areas apart from rural ones, though, is the extent to which they are designed by people, for people. Although cities obviously differ depending on where in the world they exist – shaped by different cultures, climates and their surrounding biodiversity – there are also plenty of similarities between them. We are a well-networked species, sharing information on urban design, planning and architecture through communication channels that transcend borders. Increasing globalisation of materials and technologies means that we often share the same infrastructures. There are also some predictable physical effects that come from large numbers of people living and working together, in an environment of concrete, glass and steel; most notably the urban heat island effect, which raises temperatures by up to 10 degrees in high summer compared with surrounding areas.

Whether we like it or not, we have changed our planet for good. We are the ultimate ecosystem engineers; our cities the world's new vacant niches.

GETTING THE MESSAGE

'I hope you like pigeons?' Karen asks. The purpose of our walk was to talk about foxes, but Karen has the propensity to dart from one subject to another. She's hyper-sensitive to her surroundings, eyes constantly peeled, a habit she's picked up as a wildlife rescuer. Before I can even answer, we hear an almighty whoosh as a flock of at least 100 pigeons descends from a couple of leafless trees, beating their wings as they circle around us. 'They know my jacket,' exclaims Karen, pointing at her brightly coloured parka; 'they're very clever!' She quickly ushers me off the pavement and into a scrubby little park to lead the pigeons away from the road. The flock follows us from one end of the park to the other.

Pigeons are one of just a few species that can survive in the stony heart of our cities, without soil or vegetation. Domesticated by humans for their extraordinary homing abilities, they are thought to use the sun as a compass, perhaps aided by the earth's magnetic field, as well as a sort of inbuilt mapping system – a pigeon SatNav. Scientists have tried and failed to get to the bottom of their navigational skills through a series of bizarre experiments – transporting them over long distances in blacked-out vans, attaching magnets to their legs and even forcing them to wear tiny goggles to blur their vision. Still the pigeons return home. During the First World War they played a vital role in disseminating messages from the trenches. A female pigeon called Cher Ami, donated to the US Army Signal Corps by the pigeon fanciers of Britain, famously delivered an SOS from an encircled battalion trapped behind enemy lines without food or ammunition. After her two predecessors perished in action, Cher Ami was dispatched and the Germans opened fire. She was shot down after just a few seconds, but managed to take flight again, travelling 25 miles in twenty-five minutes with a bullet through the breast, blinded in one eye and with one leg hanging only by a tendon.

Cher Ami was awarded the Croix de Guerre medal for her heroic service in helping to save the lives of 194 men, and in November 2019 she was one of the first creatures to be awarded the Animals in War and Peace Medal of Bravery, which was bestowed upon her posthumously at a ceremony on Capitol Hill in Washington DC.

In the modern world of email and text we have little need for carrier pigeons. They have thus returned to the wild, this time replacing the clifftops from where they originated with high-rises and tower blocks, becoming one of our most successful urban exploiters.

Over 200 fiery orange eyes stare up at us expectantly from the ground. Karen looks around furtively: 'You never know with pigeons whether people will be angry I'm feeding them, so I try to be discreet,' she explains in hushed tones, waiting for an elderly couple to pass by before producing a small sack of food from her bag. As she tosses the seed to the pigeons, I note how visually different they are from one another. Some are a warm chestnut colour with cream smudges, others almost entirely white and dove-like. The most common colorations are either chequered, or a classic pigeon-grey with two black stripes known as 'bars' on their back; both of these have iridescent green necks that shimmer as they catch the light.

Feeding the pigeons, Karen tells me, allows her to check for injured birds. She looks closely at their feet in particular, to see if any claws are tangled with string, human hair or other rubbish, a common pigeon complaint that can result in lost toes or, worse, entire feet.

Not everyone shares Karen's affection for pigeons. 'People used to love feeding them in Trafalgar Square – it was an experience – and you weren't told they were horrible then, were you?' she demands. 'But since that was banned people have become convinced that they are full of germs and are now scared of them, like many urban animals – rats, mice, foxes . . . And then the media makes it worse by printing lies about them.'

It's hard to deny when you look at some of the headlines. 'Fantastic? No, Mr Fox is a vicious pest' froths the

Daily Mail. 'Ring-tailed parakeets are flying beyond our control', claims *The Telegraph*. 'FLEW MURDER: Seagull attack victim says it's only a matter of time before they kill someone,' cries *The Sun*. In a particularly bloodthirsty piece in the *Financial Times*, a reader writes in to ask if it's legal to 'shoot squirrels in the city'. The *FT*'s resident property lawyer replies with detailed advice on how to stay on the right side of the law while targeting the unfortunate rodents with firearms.

City life is fraught with such tensions between humans and other species – the reality of living in ever-closer proximity. But if we're going to embrace wildness in our cities, then we also need to accept that all of these creatures are part of the diverse tapestry that makes up urban life. Once we start seeing cities not just as functional spaces where we eat, work and sleep but as living, breathing ecosystems for both people and wildlife, our lives will be all the richer for it.

As we will discover, this recognition of urban wildlife offers an opportunity to build closer connections too. We're actually more likely to encounter nature in an urban environment than a rural one. This is for two reasons: firstly, it's where the majority of us live and spend most of our time, and secondly, it's where other urban-dwelling species have grown accustomed to our presence, allowing us to get closer than we otherwise might to their more wary countryside relatives.

Some of the wildlife we find in cities was here long before humans became the dominant force and has continued to eke out a living. Other species have been introduced, both intentionally and unintentionally. And yet more have found that, when it comes to city living, the upsides just about outweigh the downsides. In this book we'll be looking at some of the more unusual forms that nature takes in an urban environment: self-seeding plants like poppies and oxeye daisies, and animals like peregrines, badgers, deer, water voles and hedgehogs – enigmatic species that I'm sure I won't have much trouble getting you to love. But I also hope that you'll come to view the more common creatures, such as foxes, frogs, pigeons, riverflies – perhaps even mosquitoes and rats – as worthy of your interest and respect too. I hope you'll consider allowing clover and dandelions their place on a lawn and grow to appreciate the odd plant sprouting through a crack in the paving. In a world where it's increasingly hard for wildlife to survive, we should make space for those species that are succeeding in an urban environment, as well as protecting the last remaining pockets of pristine wilderness that more specialised species rely on.

I'd like to take you on a journey through the diverse habitats that make up our complex, sprawling, sixth great ecosystem – the city. Rather than following the routes that humans have laid out, we'll navigate with fresh eyes, using new paths that other urban-dwelling creatures have created. Because everything changes when we start to see our cities as a series of interconnected habitats, deserving of the same attention and care as the forests, deserts and grasslands, the vast oceans and icy tundras that I learnt about as a child. Paved and polluted, congested and concreted, bright and brash, our cities are alive.

Now, let us take to the skies . . .

CHAPTER 2

Look Up: The City from Above

Once, with head as high as ever,
Johnny walked beside the river.
Johnny watch'd the swallows trying
Which was cleverest at flying.

Heinrich Hoffman, 'The Story of Johnny
Look-in-the-Air'

From above, the flat landscape of London and the south-east of England is reassuringly homely. A patchwork of fields gradually gives way to suburbia – houses, gardens, roads, schools, corner shops – increasing in density as you reach the city centre. The best way to find your bearings from the air is to look for the River Thames; that 'interminable waterway', as Marlow describes it in Conrad's *Heart of Darkness*, snaking down London's middle. There are no hills, cliffs or mountain ranges here, only tower blocks; no jungles or forests, only clusters of woodland and solitary trees, surrounded by seas of grey concrete; no wildflower meadows, only parklands, verges and rooftops.

I know this because I've spent plenty of time 'flying', as we humans describe it. Sitting inside a vast metal tube, we're far removed from the elements that truly airborne creatures face; only when our movie is interrupted by the captain warning of turbulence and switching on the 'fasten seatbelt' sign are we reminded that not everything is within our control.

But what must it be like, to really fly? That feeling of weightlessness; wings outstretched, a small, feathered body buffeted by wind and rain, heart pumping warm blood to the extremities. Some species – ravens, for instance – perform intricate aerial displays. They revel in the act of flying. Or so it seems, at least.

Up high it's not just birds and aeroplanes. Aerial plankton hangs suspended in mid-air; most of it microscopic in size – viruses, bacteria and fungi, algae, mosses and liverworts, and hundreds of species of single-celled organisms. Spiders and insects, such as aphids, are swept upwards by currents of air and found floating up to several thousand metres high, unless they're snapped up first by swifts, swallows and house martins – birds that feed on the wing.

A true bird's-eye view may be an unachievable dream for most humans, but cities do allow us to experience multiple vantage points; whether that's standing on the top floor of a high-rise looking down, or more often than not standing at ground level looking up.

Looking up has traditionally been discouraged. We're told to keep our heads out of the clouds; gazing at the sky is a frivolous activity for poets and daydreamers. My mother had an old book called *Struwwelpeter*, which I used to read as a child. Translated from the German original, it contains a collection of cautionary tales for badly behaved children, many of which end with maiming or death (no wonder it's etched in my memory). One of those tales is of Johnny Look-in-the-Air, a blond-haired boy who is forever looking up instead of paying attention to where he's walking. One day he collides with a passing hound; on another he's so engrossed in watching the swallows as they somersault above his head that he falls into a river. I imagine that if the story were written today it would be called 'Johnny Head-in-his-Phone', which to my mind is a far greater vice, of which I'm often guilty myself.

'Look up!' is also the mantra of David Lindo, otherwise known as the Urban Birder; I arrange to speak with him on FaceTime to get his advice on reaching for the sky. Sitting on my lawn at lunchtime on a warm, sunny day in mid-May, it's not the best time to be looking for birds, but I can just about make out the odd gull passing over. Every now and then a wood pigeon lands inelegantly on my bird feeder, sending my dog Fox into a frenzy. A gaggle of house sparrows briefly pop in to pick aphids off the climbing roses.

'Greenfinch!'

The sound bursts out of my mobile phone. On the other end, David is gazing up at his own patch of sky.

Growing up in Wembley in north-west London in the 1970s and 1980s, David developed an obsession with birds. As a young Black man living in a very urban environment, this was considered peculiar by his friends and family, but his mother did indulge him with his first pair of binoculars. David finally left London a few years ago and now lives in Spain in Mérida, Extremadura, in the western-central part of the Iberian Peninsula.

In normal times David is a global nomad, leading bird-watching tours around the world. Not now, though. As we speak in the midst of the first wave of the coronavirus pandemic, David's world has shrunk all the way down to his two-bed flat and a modest roof terrace. His view, which he shows me, comprises a lot of concrete and little in the way of greenery. 'I've had this apartment for two years and I've never spent much time sitting on the roof before,' he tells me, 'but now I'm up here for an hour every day around lunchtime. I've had about forty different species fly over in the last seventy days. One day I saw a golden oriole – that was quite special.'

David has developed an incredible field of peripheral vision, as well as a sixth sense for where to look. As we talk, his eyes dart around. Birds are able to compartmentalise their brains so that one half can sleep while the other stays alert. Perhaps David is the same – using one side of his brain to chat to me while the other scans his surroundings. The key is to focus on movement as much as shape, he explains; raptors soar, pigeons flap, gulls coast, swifts dart – although in practice it's rarely that simple.

During Spain's first harsh lockdown at the start of the pandemic, David found his release in the sky. 'Even an expert like me can learn a lot by looking up from one small patch,' he muses, 'and in a way, the limitations make it more interesting. For example, I knew there were bee eaters nesting in the city, but I didn't expect to see anywhere near as many as I have.'

He may not have spent much time on his own terrace until now, but David has long had a fascination with rooftops. Over a decade ago he persuaded the owners of Tower 42, a 180-metre glass skyscraper in the City of London, close to Liverpool Street Station, to allow groups of birdwatchers onto the roof. He was hoping that from this elevation he could get a closer look at the huge influx of wood pigeons that fly over from October to November – more than 40,000 at a time, according to his own estimations.

David's group of birders, armed with their state-of-the-art binoculars and telescopes, made their way through the hordes of City financiers and corporate lawyers every week in spring and autumn for nine years. It was not for the faint-hearted – some days, David admits, they would stand for hours, lashed by wind and rain, and record absolutely nothing. But then, suddenly, their patience would be rewarded with a flurry of activity: a peregrine catching a pigeon in mid-air – a moment of high drama, over almost as soon as it started in a puff of feathers.

During his time at the tower David recorded hen harriers, marsh harriers, hobbies, rooks, kittiwakes and even a honey buzzard. I ask if these sightings came as a surprise. 'Not really,' he replies with sincerity; 'things might be unexpected to many people, but me, I always expect anything.'

Whatever our physical vantage point, David suggests we try taking a leap of empathy and reimagine our urban environments from a bird's perspective. A city is not just a collection of buildings, roads and the odd green space, but a series of diverse and fragmented habitats. An avenue of trees is a line of scattered woodland which can sustain a large variety of insects, as well as the birds and other animals that feed on them. A tower block is a cliff, complete with rocky outcrops – ledges where peregrine falcons can nest. Waterbodies – the naturally occurring streams, rivers, lakes and pools and man-made canals, ponds and reservoirs, from the tiniest speck of blue to the Regent's Canal or the River Thames itself – are capable of supporting a wealth of life, as well as providing watering holes for passing migrants. And the snug space between the gutter and the roof of a terraced house might just be the perfect spot for a swift to make its nest . . .

A NICHE IN THE RAFTERS

It's an idyllic evening in late May. I've arranged to meet with Helen Shore, a volunteer for the Swift Conservation Trust, in a spot near her home in Cockfosters, where she likes to walk her dog Woody – a lovely old foxhound cross with his tail pointed upwards and nose to the ground. Technically, we're still in London, but on my drive over the streets become wider and the houses grow larger as I get further into the suburbs. Our meeting place is some way down what looks suspiciously like a country road, with national speed limit signs and fields either side. I quickly identify Helen – she's appropriately dressed for a birding expedition, with a baseball cap to shield her eyes, knee-length shorts and walking sandals. Meanwhile, I'm wearing a flowery sundress and converse trainers, which, as it turns out when I later stumble on a patch of brambles and shred the back of my knees, was not the wisest choice.

The arrival of the swifts is said to herald the beginning of spring, occurring with impressive punctuality in early May, so we've timed this correctly, although tonight it feels like summer is already upon us. Illuminating the evening with a golden glow, the sun's warm rays extend down into the ochre-hued fields. We've not been walking for long before we bump into a local member of Helen's RSPB group and his wife. They stop to chat about recent sightings; hot off the press is the news that a cuckoo has been identified in the area – for the first time, according to the local farmer, who's been working these fields for decades.

As we continue walking Helen eagerly tells me about her work with the Swift Conservation Trust. The only trouble is, there isn't a swift in sight. This is ironic, given that for the past couple of weeks I've seen swifts most evenings outside my front door as they loop through our housing estate.

As fish are to water, swifts are creatures of the air. Outside of the nesting period, the adult birds spend their entire lives on the wing, feeding on airborne insects and spiders and even sleeping mid-flight. Their long, scythe-shaped wings are suited to this high-wheeling lifestyle and their stumpy legs and feet barely serve the purpose of walking. While birds like tits and robins have evolved to be forever taking off and landing, swifts conserve the energy this requires by gliding at greater altitudes on the warm air currents, which is where you're most likely to see them. The swifts that arrive here in spring are by definition British birds, because they spend the most important part of their year in this country – the breeding season. But when autumn sets in they travel for thousands of miles to Africa in search of warmer climes and a greater abundance of insects.

It is thus all the more incredible that a bird with such an ethereal lifestyle can also be so decidedly urban. And yet such is the case. Modern-day swifts nest almost exclusively in human-made buildings, living alongside us in towns and cities, and they've been doing so as far back as Roman times, or even further perhaps, though we lack the records to prove it.

It is sadly also for this reason that swift numbers are now in decline. In order to make their nests, swifts require cracks, crevices or small cavities in the roofs of buildings – underneath tiles, beneath windowsills and ledges, in eaves and gables. This is their niche – finding and exploiting spaces that few other birds can make use of. Unlike swallows, which carefully construct their nests from scratch out of compacted mud, swifts simply pad out a suitable cavity with a few materials that they pluck from the air – bits of floating leaves, moss or feathers, bound together using their saliva.

People often confuse swifts and swallows but, although they look similar, particularly in flight, and occupy a similar niche, they're not related. Instead they are the product of 'convergent evolution' – having independently developed their distinctive traits in order to adapt to similar environments. You'll rarely see either up close, but you can tell them apart in the sky by their tails – although both are forked, the swallow's tail has distinctive streamers while the swift's is shorter and less defined. Swifts may look black against the sky but they are actually a dark, sooty brown, save for their paler throats. Swallows, on the other hand, have red throats and white undersides.

What used to be an evolutionary advantage for swifts is now proving their downfall. Unlike other small birds such as house sparrows, which can nest in all manner of nooks and crannies, when the sites that swifts need to breed are filled up or filled in, they're unable to adapt. As urban areas are redeveloped – old buildings replaced with new and lofts converted to create additional space – swifts are left with fewer places to nest and breed. This is compounded by the fact that they mate, and continue to nest, in the same spot for life.

The way in which swifts navigate back to their nest site each year remains a mystery. It has been suggested that, like pigeons, they use the earth's magnetic field, but they're thought to possess a kind of photographic memory too. If the building they usually nest in changes they will fly to the nearest spot; what a heartbreaking sight it must be when a pair of swifts return from their extensive travels, only to find their nest destroyed and scaffolding in its place. The adults will continue to fly around and around for hours before eventually giving up. Many pairs fail to breed as a result.

Fortunately, some urban-dwelling humans are trying to intervene before it's too late. Helen spends her spare time giving talks at schools and churches about the swifts' plight. Every time she sees a loft extension under way she'll knock on the door and ask the residents if they'd like a swift box installed. (The boxes need to be at least 4.5 to 5 metres high, so it's best to do this while the scaffolding is still up.) Twenty-five new boxes have been installed as a result of her efforts. She also lobbies developers to make their new-builds more swift-friendly. This can be easily done by replacing a few standard bricks at the appropriate height with so-called 'swift bricks', which include a small, swift-sized hole in the front and a cavity at the back.

But this is just the first step: the swifts still need to be able to find the newly created nest sites, and for this some advertising is required. Since swifts tend to choose an area where they know that others have successfully nested, Helen recommends placing a portable speaker nearby and broadcasting the sound of their unique joyful shrieks to draw in potential house hunters. She has been trialling this method at her own home, causing a little disquiet amongst her generally tolerant neighbours: 'They did object to the swift call early on a Sunday morning,' she admits; 'I asked my husband to turn it off and he forgot.

I got a message from next door asking "Helen, this swift call, will it be going on for days, weeks, months . . . ?"'

After walking for over an hour, Helen and I still haven't seen a single swift – although I have learnt a good deal more about them – so we decide to call it a day. I'm left determined to find another window into their urban life before they depart again for the winter, leaving the skies much emptier.

Just as we start to turn back, I hear a familiar and unmistakable sound. Not the call of a swift, but something equally welcome, taking me back to my childhood. It's faint at first, above the ever-present hum of the M25 motorway, but gradually grows in strength.

Cuckoo. Cuckoo.

My search for swifts continues a week later. The weather has taken a very different turn and the sky is grey and heavy; gusts of wind blow strands of hair across my face and into my eyes. For once I've remembered to bring my waterproof – it feels as if the heavens could open up at any minute, and yet somehow they don't. For tonight, at least, it remains dry.

I've arranged to meet Mike Priaulx, a sustainable building consultant and founder of the Islington Swift Group, at the entrance to Finsbury Park in north London, next to a Lidl supermarket. I recognise Mike straight away. Amidst the parents with buggies, Lycra-clad cyclists and stony-faced teenagers, he is the only obvious choice for a birder.

He's wearing a khaki baseball cap, with shoulder-length strawberry-blond hair and the obligatory birder beard.

It soon becomes clear that walking away from Finsbury Park is our best bet, towards the rows of terraced houses on the other side of Seven Sisters Road, where we can just about make out swift-shaped outlines darting through the muggy air. We reach a standard residential street, each yellow-brick house almost identical to the other. Trees have been planted at regular intervals along the pavement and the houses have small gardens – which in many cases seem to have been left to their own devices. It's enough to support a decent quantity of insects, but there doesn't seem to be anything special in terms of habitat here. And yet it just so happens that the way in which these houses were built, with gaps in between the gutter and the roof, makes this street an ideal place for swifts to build their nests. 'You could walk across the road to another street with a different style of houses, and you'd see virtually no swifts at all!' Mike asserts.

As if to make up for the lack of swifts during my outing to Cockfosters last week, this evening we are surrounded. The swifts fly low, immediately above our heads.

During the day swifts can travel long distances in their search for food – these birds could easily turn up in London's extremities – above Walthamstow Wetlands, Lee Valley, Epping Forest or further afield – they will forage for up to 200 miles a day, Mike tells me. Their smooth white eggs are equipped to survive without a parent's warmth for hours and the young tolerate being left for extended periods by going into a state of torpor – like a mini-hibernation.

In general a swift's preference is to stay up high, where they face few, if any predators; 'There's only the hobby they need to worry about up there,' Mike observes, pointing at the sky, as if one might be soaring above our heads at this very moment. Hobbies are birds of prey with long, pointy, swift-like wings and, as it turns out, a male has recently been spotted in Finsbury Park. 'It's not general knowledge,' Mike reveals in hushed tones, 'but he's been seen here, and also up the road in Stoke Newington. I don't really want to speculate where the nest might be, because I don't want people finding out.'

But tonight, Mike explains, the bad weather has driven the insects downwards and the swifts have followed, sticking close to their nest sites as night-time draws in. They continue to manoeuvre effortlessly between the houses, darting and dipping right in front of our noses. The speed at which they travel makes it almost impossible to study them in any kind of detail, their plumage silhouetted against the dull evening sky; I can just about make out the pale-coloured patch on their throats. With bullet-shaped, almost dinosaur-like heads, each swift looks much the same, so even for experts like Mike they're hard to monitor. He suspects that many of this evening's individuals are younger birds – over the past couple of years the weather has been relatively favourable, so numbers of young should be up, although sadly the trend is still towards a 5 per cent annual decline.

Teenage birds tend to arrive in the UK later than adult pairs and they're particularly noisy. Once swift chicks have fledged, the young birds may not land at all for three or four years, and they'll spend all summer flying around as they search for a suitable nest site. They use a technique called banging – Mike assures me this is the technical term: the young birds will bang into spots where they think there might be an appropriate cavity, in order to test it out. They also use banging as a tactic to baffle predators and distract them from the true location of their nests.

'Look!' exclaims Mike, suddenly animated. I turn just in time to catch one of the swifts bouncing off the guttering on a house opposite where we are standing. It does this a couple more times in different locations, before concluding that we lumbering predators are sufficiently confused and darting into a gap in one of the roofs. 'What's the name of the road we're on?' Mike asks urgently, recording the location of a possible nest in his notebook. He'll return later, he explains, to post a leaflet through the door informing the occupants of their new neighbours.

As we're about to leave, Mike shows me a swift box he helped one of the local residents to put up. The swifts have summarily rejected it, choosing to nest in a niche in the rafters instead. It reminds me of the pretty wooden bird box, shaped like a little house, that we put up in our garden earlier in the year. A pair of great tits briefly investigated before shunning it in favour of our neighbours' guttering. Swifts are creatures of two halves – one celestial, the other true grit.

We head back towards Finsbury Park. By the time I get home the light is fading, but just as I'm putting my key in the door I catch something out of the corner of my eye: one last swift, zipping in between my neighbours' houses before disappearing into the distance.

AN URBAN CLIFF DWELLER

We design our cities for humans, and yet wild creatures will often find a way to make use of them, especially when the structures we create happen to resemble their preferred habitat. This is sometimes referred to as 'preadaptation' – the relatives of modern-day swifts used to nest in trees, but they found a similar niche (in both senses of the word) in urban environments. Over thousands of years swifts have evolved to use our buildings to raise their young, and no longer nest in trees at all.

There are other high-flying species that are similarly preadapted to the urban environments that we build, but some have made their move to the city much more recently.

The peregrine falcon is famed for being the fastest creature on earth when making its gravity-assisted dive – too fast, it seems, for my reference guide to keep up: 'breeds on steep coastal cliffs or in mountains', the relevant entry reads.

As I jump off the Tube at Holloway Road, just a few stops from where I live, there isn't a clifftop or mountain in sight. I'm here to meet with Stuart Harrington of the London Peregrine Partnership, who is waiting for me just across the road from the station. Stuart is a careful-looking man, with glasses and slightly greying hair. He works as a Web developer, running the London Peregrine Partnership in his spare time alongside a small team of volunteers. Their aim is to protect London's peregrines and ensure their breeding success.

We stand at the foot of an unappealing 1960s high-rise, owned by London Metropolitan University. It's made up of two buildings stuck together – the front building a dirty grey colour with rows of blue-tinged windows and behind it a taller concrete structure. That's what it might look like to my eyes, but to a peregrine it is for all intents and purposes a cliff. It has the necessary height, between fifteen and forty floors, and, like a real cliff, it's furnished with plenty of ledges and spots to perch on all the way round. When the peregrines first arrived the only thing this particular site lacked to help with the nest-building process was gravel, which at Stuart's request the university obligingly provided.

I hear the harsh cry of the baby peregrines almost immediately – before Stuart even has the chance to show me where the nest is located. We spot the mother first, perched on a railing at the top of the taller building. Her compact, crow-sized body belies the peregrine's fearsome reputation. She is at once perfectly still and fully alert as she gazes out, surveying her city.

I've only ever seen a peregrine once before: a few years ago on a clifftop at Beachy Head, just outside Eastbourne on England's South Coast. It was a dreary day and its bright yellow beak, legs and feet stood out against a backdrop of white and grey cliffs and murky waves. The colour palette in this concrete landscape isn't dissimilar, but seeing a bird of prey in an urban environment feels somehow different. When I tell a friend I went to see peregrines in Holloway she assumes the birds have been deliberately introduced to deter pigeons, confusing them with the Harris's hawks used by pest controllers. To many city-dwelling humans, urban peregrines are as unknown as they are unexpected.

The female remains motionless, but out of the corner of my eye I spot a flash of movement – it's her partner, heading off to hunt. Stuart is familiar with this pair – they were ringed as chicks, making it easy to keep tabs on them. They nest in the same spot every year but, slightly unusually, they have another roost over the winter months on the roof of St Pancras Station.

Stuart's fascination with these birds began as a child. He remembers owning a field guide with colourful illustrations of different raptors. 'They looked so majestic,' he recalls. 'I guess the peregrine wasn't any more majestic than the other birds of prey, but it had this added mystique about it because it was so rare at the time.'

In the 1960s peregrine numbers dipped to such perilous levels that the species almost became extinct in the UK. While populations did gradually start to recover, it wasn't until the late 1990s and early noughties that peregrines began to take to city life. The first few London pairs may have come up from the South Coast in search of new nesting sites, drawn in by the Thames, which acts as a magnet to so many species. By the time Stuart moved to the capital as a student in the 1990s there were peregrines nesting in Battersea Power Station and on top of the Tate Modern. He got in touch with a couple of locals who were monitoring them at the time, and next thing he knew he was spending most evenings and weekends visiting nest sites.

The peregrines that made their way to urban areas soon found plenty of pigeons to eat and strong conditions for hunting. Cities like London form urban heat islands. During the daytime large sections of concrete warm up, creating thermal currents which are perfect for coasting above the tops of buildings in search of food. Pigeons might be the marathon runners of the bird world, but peregrines are sprinters. They gain an advantage by flushing their prey out into the open – above a body of water, a green or a common. Only once they've reached enough height, and the timing is right, will they attempt their 'stoop', descending at speeds of up to 200 miles per hour, beak first, and hitting their target with incredible precision. The force can be powerful enough to decapitate a pigeon in mid-air.

Most peregrine kills are a little less dramatic, though. 'What you generally see is more like a mugging,' Stuart claims; 'when a less streetwise pigeon finds itself flying too high or out in the open it's easy for a peregrine to sneak up in its blind spot. It's all over before the pigeon even sees it coming.'

As if on cue, Stuart and I are interrupted by a dismembered pigeon wing tumbling from the building behind us. The culprit, a crow, peers over the side suspiciously. Crows will often hang around peregrine nest sites, Stuart explains, on the hunt for leftovers. If they're feeling particularly foolhardy they might even gang up on one of the adult peregrines, hoping that it will drop its prey. But these particular crows have sussed out an easier way to find food, having learnt that the Holloway peregrine family stores some provisions on an adjacent building. Once the peregrines have headed out for the day the crows sneak in and raid their larder.

While pigeons form their staple diet, metropolitan peregrines have diverse tastes and will opt for something more exotic every once in a while. Rows of little red beaks have been found: gruesome evidence that London's most colourful coloniser – the ring-necked parakeet – is a recent addition to the menu. Stuart has also found leftovers to include redwings, fieldfare, woodcock and lapwings. 'I was at King's Cross one morning and I saw these feathers coming down,' he recalls sombrely, 'and it was obvious that they were being plucked from a prey item. I picked them up and discovered the remains of a golden plover. We've even found feathers from nightjars and cuckoos.' A peregrine is unlikely to have picked up a golden plover in King's Cross, but the hunter may have recently returned

from Woodberry Wetlands or Hampstead Heath in the
north of the city, or the London Wetlands Centre in the
south, where many migrant species pass over, especially
in winter.

In the time that Stuart and I have been talking only
the female peregrine has remained visible, unmoved on
her sentry post. But suddenly a pair of wings appears on
the ledge below her, flapping in an urgent and ungainly
fashion. A head emerges – it's one of her young – and
soon after another sibling pops up alongside it. 'They're
building up muscles,' Stuart explains, about the flapping.
'It's like doing press-ups. They'll be ready to make their
first flight any day now.'

Adult peregrines don't face any real predators here in
the UK – even less so in the city – but these fledglings are
vulnerable; many will live fast and die young. As they
take their first few flights the young birds can crash-land,
exposing them to oncoming traffic and a life-or-death
struggle to make it back up to the nest site. Modern
skyscrapers made mostly of glass are an added peril. As
the sun reflects off the windows the young birds find it
hard to distinguish between glass and sky; the resulting
collisions can be fatal. The netting many caretakers add
to their buildings to deter pigeons also creates issues –
Stuart recalls a particularly harrowing incident in which
he was called to check on a young peregrine, only to find
it hopelessly entangled and mummified in a net.

It takes time and dedication for young peregrines to
perfect their hunting technique, and the Holloway fledg-

lings have a long road ahead. They'll start with short practice flights and once they're more seasoned at flying the siblings will begin play-chasing – first one another, then other birds – a bit like kittens – endearing, until you remember what the training is for. The parents are actively involved in this learning process. Like the Deliveroo riders that beetle about the streets below, they deliver food directly into gaping mouths while the chicks are bald and helpless, but once the young birds start to take flight the adults drop morsels from above, encouraging them to work for their dinner.

Gradually the fledglings become more independent, heading off to explore during the daytime, before returning in the evening to roost. By the end of the summer the adults are pretty much ready to see the back of them. The male offspring will stay nearby, but the females usually venture further afield, living up to the origins of their name (a 'peregrination' being a long and meandering journey).

It's nearly time for me to venture further afield too, but Stuart is a little anxious – he's not visited this site for a while, and though he knows that three chicks hatched, we've only spotted two. He suggests we walk to the other side of the building to see if we can get a better view from there. Just as we're about to do so, another pair of wings emerges and he breathes a sigh of relief. Now that he has chicks of his own, Stuart doesn't have as much time to make site visits, so he focuses on engaging and organising others. 'In a city the size of London, only a handful of people get actively involved in peregrine conservation,' he observes with slight disappointment, 'and some days I think, should this have been something that I pursued full time? Would I be doing what I'm doing now?'

A few days later I return to the nest site alone – I want to see if the young peregrines have taken their first flight – but I can't make out any sign of them or their parents. I spend a few minutes searching from the Holloway Road with my binoculars (and get some strange looks from passers-by in the process). Just as I'm about to leave – I'm running late to meet a friend – I catch two shadows in my peripheral vision. They land on one of the many ledges, before quickly disappearing out of sight. It could be the adults, but I'd like to think it's the fledglings, returning from their maiden voyage.

UP HIGH IN THE LOWLANDS

As spring turns to summer, my partner Ben and I plan a trip to Scotland. Ben likes to spend his annual leave walking off-grid, so we come up with a compromise and drive to Edinburgh for a couple of nights first, before continuing on to the Highlands and the Isle of Skye. Despite his desire for remoteness, Ben is quickly as enamoured with Edinburgh as I am. While enjoying a G&T in a bar with views of the castle, he suggests we think about moving here permanently 'if everything falls to shit back home'.

The next day I'm up and out at 8.30 a.m., ready to

meet George Smith, of the Scottish Raptor Study Group. George has suggested we convene 'at 09:00 at the Broad Pavement Car Park near Holyrood House, Grid Ref NT270737'. The car park is next to Holyrood Park, where we are to begin our search of the Edinburgh skyline for peregrines, kestrels, sparrowhawks and buzzards – the Scottish capital has them all – plus a pair of cliff-dwelling ravens, believed to be the first of their kind to move into this city.

The word raptor comes from the Latin *rapere*, meaning to seize or plunder; once airborne, these birds are the undisputed masters and mistresses of the sky, with piercing eyes, powerful beaks and sharp talons that hark back to ancient reptilian relatives. When they're not riding the thermals they nest and roost in high places, of which Edinburgh has plenty.

The walk from my hotel on Princes Street in the city centre takes less than twenty minutes. Since I have the word 'park' in my head I'm looking for a large, grassy area with some flower beds or decorative borders, maybe a water feature or two. I reach the end of a residential street, head buried in my phone while I consult Google Maps and reply to a message from a friend. When I finally look up, I'm confronted by an extinct volcano, framed by hills, lochs, glens and ridges, its sheer basalt cliff face looming large in front of the Scottish Parliament building. Arthur's Seat, wrote Robert Louis Stevenson, is 'a hill for magnitude, a mountain in virtue of its bold design'. I wonder if it's cheating to look for urban wildlife here, and yet we're not even a mile from the city centre.

Once again I'm inappropriately dressed; having anti-cipated a relaxing stroll around the city I reached for my Converse and left my walking boots at the hotel. George – a friendly looking man in his early sixties with kind eyes and rather more hair on his chin than his head – glances down at my feet anxiously. 'I've done tougher hikes in flip-flops,' I reassure him.

Despite being almost double my age, George marches up the hill at an impressive speed while asking me various questions about myself, which I'm unable to answer – I can just about keep up with his pace, but I definitely can't chat at the same time. George works as a facilities manager for a large banking group, he tells me, using his spare time to drive across the length and breadth of south-east Scotland, all the way up to the Borders, track-ing raptors. What started out as a passion has become an obsession, and during springtime he spends countless waking hours ringing peregrine and buzzard chicks, as well as adult birds. Ringing allows him, and anyone else who is studying these birds, to gather information on their comings and goings. He recently discovered a ringed female peregrine nesting on a nuclear power station at Torness, East Lothian, and was able to trace her origins all the way back to another power station – Battersea – 363 miles away in south London.

He draws my attention to a peregrine nest site where he ringed some chicks earlier in the year, on a ledge on the other side of the hill, halfway down a sheer cliff face. This kind of complication is no deterrent for George, although he admits that he once fell off a cliff, breaking three ribs. 'The next day the doctor phoned me and said, "We've just checked your X-rays and you may have a cracked vertebra,"' he adds, 'but it was too late – I was already out on another cliff.'

While George recognises that abseiling down rocky precipices isn't something he'll be able to do indefinitely, he has yet to find anyone to take on the mantle – most volunteers don't even last a season. 'If I can do it, surely anyone can,' he protests, clearly underestimating his own stamina.

August isn't the best time of year to go out looking for raptors, George warns me – the young have nearly fledged and are likely to be off exploring for much of the day while their parents hunt. But he needn't have worried; we're less than halfway into our climb when, as luck would have it, a female kestrel appears. Unusual amongst raptors, kestrels are dimorphic – the females and males can be differentiated by colour; while both have light-brown feathers with dark spots, the males have blue-grey heads and the females are brown all over.

The kestrel hovers, her pointed wings bobbing slightly while her head remains still, as if she's suspended by an invisible thread, like a mobile hanging in a child's bedroom. She's scanning for mice and voles in the long grass below, George explains. I try to imagine what it must be like to have such powerful eyesight, every blade of grass rendered in incredible detail.

This characteristic hover is a helpful way to tell kestrels apart from some of their raptor relations, George tells me. Kestrels hover. Sparrowhawks glide on broad, blunted wings. 'And peregrines look like, well, peregrines!' he claims, as if it's the most obvious thing in the world. 'It drives my kids mad,' he admits, having noted my dubious expression; 'I'll see a shadow and say, "Oh there's a kestrel, there's a buzzard, there's a peregrine" – I can just tell by the way they fly.'

This particular kestrel, perhaps growing tired of hovering, lands on a stumpy tree to the left of our path – she seems unfazed by our presence. Just over the hills, the non-urban kestrels tend to be much more flighty, George observes. We're still watching her, when all of a sudden we're interrupted by a very strange sound – a deep, burbling, toad-like croak.

'Raven!' George shouts excitedly.

Off the side of the hill we catch sight of one half of the only raven pair known to nest in central Edinburgh. They've been here for just five years, with the next closest pair way off in the hills at the edge of the horizon.

Despite his primary focus on birds of prey, George has been busy ringing ravens too, which is relatively easy to combine with peregrines, both species opting for similar cliff and mountain-based nest sites. 'My predecessor hated them,' he adds; 'I'm not sure why, but he couldn't understand why I bothered with them – big black brutes he called them. He was anti-religion, anti-English, anti-ravens. Nice man though, if he liked you.'

Ravens are the largest members of the corvid family and, unlike their smaller crow cousins, they use the air currents as raptors do, to glide. We watch as this particular individual (even George can't tell whether it's the female or the male) soars alongside the cliff edge without flapping its wings even once. Ravens are fond of showing off, performing synchronised dives, flips and somersaults during their intricate mating displays. They mate for life, and this aerial dance is thought to strengthen their bond. According to George they have distinct personalities too. 'I've been out trying to ring chicks and the parents will scold me, ripping up bits of grass and pelting me with them,' he tells me.

Sadly, both ravens and raptors are at risk of persecution in their native habitats from gamekeepers, who see them as a threat to the grouse chicks they breed for shooting. Killing these birds is a criminal offence, but it's hard to police. To urbanites like me, driven grouse-shooting (where the grouse are flushed out of the undergrowth to be shot) is a strange and antiquated pastime, which is why cities can provide a place of sanctuary for these unfairly maligned avian predators.

As the raven disappears over the edge of the cliff George and I walk the short distance left to the top of Arthur's Seat. It's a clear, sunny day and from this vantage point we can see across most of Edinburgh. A bird of prey flying overhead will find water on one side, rooftops, spires and hills on the other. George gestures towards various nesting sites – sparrowhawks in the cemetery, peregrines in the church, buzzards in the hospital grounds. I'm struck by how green this city is – dotted with mature trees, which offer cover for birds and bats, and networked with green corridors, providing pathways for terrestrial species. Foxes, badgers, hedgehogs and deer roam all the way into the city centre, George says.

We're just about to head back down the hill when George, ever alert, spots something over on the other side. We catch a glimpse of a sparrowhawk as it drops down to roof level, and then even lower, steering effortlessly through rows of houses less than a metre or two off the ground. This is a form of stealth mode, allowing the bird to stay hidden from unsuspecting prey. Flying at speeds of up to 45 miles per hour requires lightning-fast reactions, leaving them just hundredths of a second to dodge any obstructions in their path.

Edinburgh is full of sparrowhawk-friendly habitats. Having evolved in patchy woodlands, sparrowhawks are perfectly at home in the city's larger gardens, with bird feeders providing the bait to lure in their prey. There are also plenty of overgrown cemeteries, offering a good mix of cover, open spaces for hunting and tall trees for nesting. George warns against straying into these cemeteries alone, however, sharing a cautionary tale. A friend and fellow raptor enthusiast was high up in a tree ringing sparrow-hawk chicks, when a couple below began engaging in amorous activities. Uncertain how to explain what he was doing up in the tree, George's pal opted to stay put for two more hours, until he felt sure that the coast was clear.

Holyrood Park having exceeded all my raptor-based expectations, George and I head off to our next stop – Barclay Viewforth Church – which we've just seen from above. A couple of peregrines have made their home in the church's 75-metre spire, and are often found perching on the arches, like a pair of avian gargoyles. In front of the church is a large green, providing a ready-made hunting ground.

The pair were first brought to George's attention by local churchgoers, who got in touch when a macabre confetti of bloodstained feathers, beaks and claws began raining down on the congregation.

Unfortunately, there's no sign of the peregrines today, save for a few forlorn pigeon feathers at the church entrance, so we jump back into George's car and make our way out to the leafy suburbs of south-west Edinburgh. It's only a short drive, but George is keen to justify why our final destination, a disused quarry nestled beneath the Pentland Hills, can still be considered 'urban'. Firstly, he argues, it's within the city limits. And secondly, the peregrines nesting in this quarry use the city centre as their hunting ground.

We're greeted by a number of 'Keep Out' signs, but George assures me that he has permission to enter. Crunching through a patch of gravel and sawn-off tree trunks, we reach a wire fence, beyond which is a large and rather beautiful pond containing smooth and palmate newts. Behind the pond looms a large exposure of igneous rock, formed from layers of lava, embedded with volcanic ash 410 million years ago. Our arrival disturbs a group of young roe deer with tall, pricked ears and big, nervous eyes. On seeing us they scatter on fragile-looking legs.

The scene might be quite peaceful, if it wasn't for the peregrines. One of the young birds is making a real racket as it demands with stubborn persistence to be fed. 'I wish they wouldn't advertise themselves like this,' George sighs, explaining that because the quarry is easy to find and relatively accessible, they've had a number of chicks stolen over the years. There's money to be made in selling the birds for falconry – the thieves will either use them for breeding or sneak them out of the country. At one point it got so bad, George says, that they had a team of volunteers covering the site with a twenty-four-hour watch to keep the peregrine-nappers at bay.

Considering the noise it's making, it takes us a surprisingly long time to spot the young bird that's the source of the din, perched high on a ledge. And today it seems that peregrines are a bit like buses, because no sooner have we identified one than a second appears, flying in from the left in hot pursuit of a pigeon. The two birds speed past us as the pigeon's life hangs in the balance. We hold our breath and the peregrine prepares to make its stoop. In a flash it drops down, but it has misjudged. The pigeon makes a lucky escape.

'Oh my! He's lost it!' George exclaims, visibly excited.

Out in the distance three more silhouettes appear – the mother, father and another sibling – the whole family reunited. It's a fitting end to my Edinburgh peregrinations.

TOWERING ABOVE

It's a sunny autumn day, and with the changing of the clocks trees across London have been marshalled into new seasonal hues of yellow, orange and brown. I've started spotting jays again on roofs across our estate, their floppy flight giving the impression of slight inebriation. Yesterday my neighbour Matt messaged me a grainy photo of one such corvid, perched on a peanut feeder shaped like a miniature garden bench. With curious eyes, black moustaches and blue chequered wing feathers – often favoured by milliners – jays are the modern-day dandies of the bird world. I recently read that these birds are thought to have been responsible for turning two abandoned fields next to Monks Wood, a nature reserve in Cambridgeshire, into new woodlands, thanks to their penchant for burying acorns. Being relatively shy, they become more visible in urban gardens as food becomes scarcer, and people often think they've spotted something exotic – Matt is surprised when I tell him that his flamboyant-feathered visitor is a member of the humble crow family.

Today, though, it's not jays that I'm looking for; instead, following my brief initiation in Edinburgh, I'm hoping to get closer to the largest member of the corvid clan.

I haven't been to the Tower of London since I was a child, when my mother and I queued for what felt like a very long time to peer into a glass cabinet full of shiny metal, polished stones and dead stoat fur. I arrive late and flustered – it's an unseasonable 18 degrees in the sunshine and I made the mistake of wrapping up in a winter coat, hat and scarf this morning. I'm here to meet with Christopher Skaife, better known by his official title, the Ravenmaster ('It means absolutely nothing, except that I care for the ravens,' he jokes).

While tourists still list the Crown jewels as their number-one reason for visiting the Tower, followed by the Beefeater guards, the resident ravens take a respectable third place, Christopher tells me with pride. He's a jolly-looking man with a greying beard, dressed in full Beefeater get-up himself, complete with royal insignia and a huge round hat. I explain to him that after my brief introduction in Edinburgh, I'm keen to learn more about these cliff-dwelling crows, and few people have a more intimate relationship with them than he does.

In fact, the relationship between humans and ravens is a long-standing one; over thousands of years these birds learnt to live alongside us, perched on top of buildings or soaring above our towns and cities, until they were actively forced out. Only now are ravens beginning to return to urban outskirts and suburbs, although they haven't yet reached the city that I call home – the nearest breeding pair is around 30 miles from the Tower, as the crow flies.

Legend has it that if the ravens leave the Tower of London then the kingdom will fall. This legend was most likely concocted by superstitious Victorians, but that doesn't seem to have stopped it from taking hold. People love to romanticise about the ravens – it's been claimed that during the Blitz they were used as unofficial spotters for enemy bombs and airplanes. During this period all but one, an individual named Gripp, died from either the bombing or the stress. Fortunately, Britain survived both the Second World War and the loss of the ravens, which were replaced by Churchill after the fighting ended.

There are further myths surrounding how, and when, ravens were first brought to the Tower of London. It's rumoured that the practice started under the rule of Charles II in the seventeenth century, but there's suspiciously little evidence for this. What we do know is that centuries ago ravens were a familiar sight in cities across Britain and, whether with human encouragement or not, they probably were present in and around the Tower. Ravens are thought to have learnt over the course of thousands of years to follow human hunter-gatherers, swooping in on our leftovers like house sparrows did. They stuck around as we formed settlements, and people bestowed upon them a mystical significance that persists to this day ('I've got a lot of goth followers on social media,' Christopher quips). Up until at least the sixteenth century ravens were welcomed in towns and cities as unofficial bin men, clearing the streets of offal and other unsavoury deposits through their scavenging. But over time their numbers waned, and the last wild pair in London were recorded in Hyde Park in 1826.

Christopher often gets reports of possible raven sightings in the capital, but they are largely erroneous. 'One or two could be ravens passing over,' he tells me, 'but mainly it's just large carrion crows.' The best way to differentiate them is by getting a clear view of their size (ravens really are much larger than other members of the crow family), watching to see if they coast and soar, or listening out for their distinctive croak, he adds.

Why did the ravens leave our cities? Experts suggest that various factors were probably at play: improved sanitation and the clean-up of the streets left fewer feeding opportunities, while intensive agriculture in the countryside provided grains to snack on and a better chance of a meal outside the city. Persecution probably also played a part. 'Unfortunately, they're not everybody's cup of tea,' Christopher laments; 'They're slow at moving into territories, but very quick at latching on to human ideas, and then humans kill them. I'm not going to go too deeply into the politics of it . . .'

Christopher and I head for the raven enclosures, making a brief pit stop to pick up some biscuits which he stashes in a large pocket ('Poppy will bite my feet if I go up there without any biscuits,' he adds by way of explanation). En route, I'm introduced to some of the Tower's regular visitors; 'Hi Ronnie, hi Reggie!' Christopher greets a pair of magpies, 'Oh, here's Cyril!' A squirrel cuts across the path in front of us and leaps towards Christopher, grabbing hold of his coat and scaling up to his outstretched palm, where it snatches a biscuit in its tiny hands.

Since he began his role at the Tower in 2011 Christopher has been making some sweeping changes. He's ex-military, he tells me, having joined the British Army when he was just eighteen, but nature has always been his first love. 'My boss is David Attenborough,' he maintains earnestly, 'that's who I answer to.' After taking on the role of Ravenmaster, his first move was to ban fox-culling in the grounds; the Tower has lost a few ravens to foxes over the years, but Christopher believes a balance can be struck, allowing both species to coexist. He's also made efforts to encourage and protect the four species of bats that have been recorded at the Tower, which roost under the arches and in the roofs of the old houses that some of the staff live in during the week. His latest pet project is to replace part of the grassy moat with a wildflower meadow, the same moat which in 2014 became home to a temporary installation of 888,246 ceramic poppies, 'planted' in memory of those who died in the First World War.

Despite his close relationship with the local wildlife, Christopher has been trying to put some distance between himself and the ravens. He leaves their enclosures open and allows them to choose where to roost for the night. Previous Ravenmasters used to clip the birds' flight feathers to stop them from leaving the Tower. Christopher put an end to this practice, allowing the ravens to come and go as they please.

As we approach the empty enclosures the ravens are amusing themselves on the crest of a grassy bank. A couple stand on the remains of an old stone wall; another struts along a metal perch. Their jet-black feathers reflect the warm autumn light with a most beautiful sheen. Up close I can finally appreciate their size – with an almighty wingspan of 1.5 metres they are far, far bigger than a common carrion crow; or a rook or a jackdaw for that matter.

The ravens are clearly roused by Christopher's presence – they begin making a lot of noise, their deep, guttural croaks like the grunting of an old boar. While some are warier of him than others, at least a couple have become imprinted, Christopher admits, viewing him as a pseudo parent figure. He pulls out a hose and fills a large bird bath with water – one bird tugs at the hose while another splashes gleefully. Their behaviour is almost dog-like – two more sidle up to us, heads tilted, eyes fixed on Christopher's pocket. A female named Poppy pulls at his coat insistently until he dispenses a biscuit, which she skilfully catches in her beak. Another bird, Georgie, jabs at my shoe and pulls my tights. Christopher affectionately shoos her away: 'You've had enough!' She looks up indignantly, beak still full of biscuits. She'll cache them away in her enclosure before coming back for more, Christopher says.

Academics from Queen Mary University who studied these ravens have found that they're particularly good at playing games. 'We created a version of Kerplunk using spaghetti,' Christopher tells me, 'and an old female of ours was able to complete the game in forty-five seconds.' Some, though not all, ravens are able to recognise their own reflection in a mirror, Christopher claims, and he's convinced that they are even capable of empathy. He

72

describes a female at the Tower who was in a same-sex pairing (not uncommon, apparently). When her partner died she fell into a deep depression. 'I'm not an ornithologist or a scientist,' he notes, 'but I have spent an unhealthy amount of time watching ravens, and I've seen the whole range of human emotions in them – love, sadness, joy . . .' He pauses. 'Except hate – I've never seen hatred. Only survival. They are incredibly good hunters.'

For the most part ravens are opportunists that are happy scavenging, but when they do hunt it can be pretty brutal. Christopher compares them to the velociraptors in the *Jurassic Park* movies; working together, one will chase a pigeon towards an accomplice, lying in wait, who will seize the unfortunate prey and rip it to pieces. 'I've seen two ravens eat a pigeon completely, including its feathers, in three minutes,' he claims, 'and Merlina here I've seen take a bluetit in mid-flight.'

While I could stand and watch the ravens all day, Christopher needs to get ready to give a talk to members of the public. He takes this part of his role incredibly seriously, educating people about the ravens and about wildlife in general. He is laying the groundwork, he hopes, for ravens to receive a warmer welcome when they finally reclaim their territories in cities across the country where they have long been missing.

The good news is that ravens are once again making urban inroads, particularly in the west and the north of the British Isles. They are finally starting to expand eastwards too, breeding at a handful of sites just outside the London boundary. Like peregrines, some are using the ledges, niches and precipices that we've built into our urban structures as locations for their roosts and nests. Hopefully it won't be long before they're living amongst us again.

STREETS IN THE SKY

On 20 March 2020 nearly three-quarters of the residents of the Cambridge Road council estate in south-west London voted for its demolition. Under the new plans, its 820 homes will be replaced with 2,000 flats, to be built by the rather ironically named Countryside Properties. The local Lib Dem-led council welcomed the decision from a heavily deprived pocket of otherwise affluent Kingston upon Thames.

Cambridge Road was previously used as a backdrop for crime dramas, including *The Bill* and the 2019 BBC series *The Capture*, but real-life Chief Superintendent Sally Benatar, Commander for South-West London, said she was 'relieved' at the outcome of the ballot. Local police added that the removal of the estate's many alleyways, which they claim provide hiding places for criminals seeking to evade capture, will create a safer neighbourhood.

But not everyone was celebrating the decision. Ecologist and bat specialist Alison Fure, who has been monitoring the way in which bats use the estate for over ten years, began surveying the area with increased intensity. She's been working hard to engage residents with the bats and other wildlife that inhabits the estate for some time, she tells me, through conducting 'sound walks', focused on 'treasuring the sounds around us', while a colleague of hers has been writing poems about the estate's many trees.

I meet Alison in front of Kingston Cemetery, which borders the estate on its south side, at dusk on a late-September evening. She is delightfully eccentric, with a mop of wild wavy hair and an infectious energy. She tells me that while most people claim that the bats live out their bat-like lives in the cemetery behind us, she has garnered a wealth of evidence to prove otherwise. They are dependent on the estate for foraging, she explains.

Of course, the estate wasn't designed with bats in mind; in fact I would venture that it's one of the last places people would expect to be a source of biodiversity. Its 1960s design, Alison explains, was based along Radburn design principals, a type of housing initially developed in Radburn, New Jersey in the US, which was itself inspired by the British garden city movement. It's typified by its landscaping, including plenty of native tree species, and by the now much-maligned alleyways and pathways that run between the houses like arteries, connecting each portion of the estate. (In my own 1980s housing estate, similar alleyways have long been blocked off to deter drug-dealing, which my neighbours tell me was a common problem until relatively recently.) Alison, who has clearly developed a deep love for the estate, refers to these as 'twittens', an old Sussex dialect word for a narrow path or passage between two walls or hedges. The idea of these twittens was to separate people from traffic, providing a convenient way to get around on foot. The estate also features 'streets in the sky', first pioneered by the modernist architects Alison and Peter Smithson in the 1950s. These aerial walkways were designed to allow people to meet and get to know their neighbours, safely away from busy roads.

This is one of the last days of the year that we're guaranteed a good number of bats – 'the last knockings', as Alison describes it; 'We don't go out looking for bats below 10 degrees,' she adds. In this period, just before the bats go into a form of hibernation, they are, bizarrely, mating. They use a delayed method of implantation and suspended ovulation, where the females will only become implanted with the sperm in the spring when the temperature once again rises. During this mating period the males will begin to ooze with a sex hormone produced from glands in their cheeks – highly attractive to female bats, I'm sure. They will mate with as many females as they can in small transitional roosts of around two to four bats. Soon afterwards, the bats will go into torpor somewhere close by – a tree in the cemetery perhaps. It's not a full hibernation; they will wake up every now and again to drink, or grab a snack if the weather is warm enough for any insects to come out. 'I go out like a crazy woman all through the year,' Alison jokes; 'last year, I recorded bats on the nearby Hogsmill River all through the winter. If you think like a bat you can usually find them.'

Alison's love of bats began through birdwatching, when she would encounter the odd bat in daylight hours as they were coming out of hibernation; 'And to be honest,' she confesses, 'my eyesight isn't as good as it used to be for looking through binoculars, so looking at birds is too difficult these days.' As we step towards the outer edge of the estate she takes out two bat detectors, which use the sound waves that bats produce when they echolocate to determine their whereabouts. The first is a tiny square device that plugs into the bottom of an iPad. These can be purchased for £300, or £195 for a non-professional version, which offers very similar functionality, according to Alison. Through an app on the iPad a sonogram – an image of the sound waves – pops up on-screen when a bat is detected, along with the name of the species. It can even generate a map of any sightings.

Alison hands me a more conventional bat detector, which looks like a small portable radio. She uses the dial to tune it to a frequency of 45 kilohertz. Each bat species echolocates at a different frequency, she explains, and 45 kilohertz is the frequency used by the UK's most common species, the aptly named common pipistrelle. In addition to common pipistrelles, Alison regularly finds soprano pipistrelles here on the estate too. Until 1999 the two were considered to be the same species. They look almost identical, but the sopranos are slightly smaller in size and, as their name suggests, use a higher frequency of 55 kilohertz to echolocate.

'If you're on a species' peak frequency and an individual shows up, your bat detector will make a clicky sound,' Alison explains, 'but if you're off, it sounds tinny, and then you need to retune.'

Occasionally, Alison picks up the Nathusius' pipistrelle too, a Continental migrant that uses a frequency of 39 kilohertz (it's impressively precise). She's also recorded high-flying noctule bats, a far larger species that uses a much lower frequency of 20 kilohertz and can therefore be detected at greater distances of up to 20 metres. Unlike the common and soprano pipistrelles, which only travel a couple of kilometres a night and are therefore dependent on the estate for their evening forages, noctules can travel long distances. Alison guesses they might be flying over from Richmond Park towards the Thames, using the nearby Hogsmill River as a corridor. They used to fly together with swifts, she tells me; 'but these days you can't even find any bloody swifts'.

While we're still talking, my bat detector starts clicking furiously; Alison confirms that it's a common pipistrelle. Looking up, we catch its silhouette as it flits above our heads. It's rarely possible to see these winged mammals in any great detail as they dash through the dark air, but up close they are strange-looking little creatures. Not so long ago Ben's stepmum Suzanne found one in her house, clinging to a doorway. She sent us a photo. It was mousy-coloured and fuzzy, with huge ears, tiny eyes and a broad nose, its black leathery wings folded at its sides like strange arms with pointy elbows.

In this part of the estate there's just enough tree cover, and it's just about dark enough for the common pipistrelles to forage successfully. Bats are not blind, despite the common expression, but like us they can't see in the dark using their eyes alone. In order to navigate and to 'see' at night-time, they need surfaces that their echolocation calls can bounce off, so they avoid open spaces. The tree lines of the estate make the perfect navigational aid, as well as creating a shield around the streetlights.

While the frequencies that bats use to echolocate are far beyond the reach of our hearing, the calls they make when communicating with one another can sometimes be heard by younger people, who haven't yet lost that range. 'When I do bat walks,' Alison tells me, 'children sometimes put their hands over their ears and say, "Oh Mummy, my ears are hurting!" The adults think they're being ridiculous, but the kids can hear the bats chittering.' (It makes me think of Ben, who recently discovered he is no longer able to hear the high-pitched sound of grasshoppers or crickets. I thought he was having me on at first, which is exactly the kind of thing he'd do, but no – aged thirty-three he's already lost that register in his hearing.) Later on in the evening, after the bats have had a feed, the social calls become more frequent. 'Really, they should be called antisocial calls a lot of the time,' Alison jokes; 'if you've got a common pipistrelle flying around a territory looking for a mate, particularly at this time of year, then they are mostly seeing other bats off. It sounds like a "chonk chonk chonk" on the bat detector.'

The final sound it's common to pick up on a bat detector is a 'feeding buzz', which resembles someone blowing a raspberry. Most bat species eat different types of prey, so communities of native bats can live in close proximity without interfering or competing with one another. Pipistrelles feed on clouds of small flies and mosquitoes, mainly at dusk, but if it's warm enough for the insects then they can stay out all night. The Cambridge Road estate provides an abundance of such insects, an essential component of any ecosystem, thanks to its diversity of native tree species, which flower at different times throughout the year. Well into the summer its many limes come into bloom long after the other trees have finished flowering; in June and July, Alison tells me, the canopy is full of black and orange painted lady butterflies sipping nectar from the linden flowers. Cambridge Road is also home to a colony of 150 sparrows, and regular visitors include jackdaws, crows, parakeets, grasshoppers and cinnabar moths, with dayglo red and black wings.

In total, there are 183 trees on the council's list, many of which will be felled, 'and they will replace them with ornamental species – little lollipop street trees that don't need any maintenance but do nothing for wildlife', Alison derides. The further fifty trees that currently grow in private gardens will not be replaced at all.

We move on further into the estate and make our way down one of the infamous twittens. Alison warns me to watch my step. Dense foliage spills over the fences from the gardens on either side; residents have planted apples, plums, cherries and even an olive tree, attracting yet more insects. Broken soffits and slipped tiles on the roofs of the houses provide warm, undisturbed spots for bats to crawl in and roost and sparrows to nest. This is clearly one of the bats' favourite haunts – here they've made their own 'streets in the sky' as they use the alleyways like aerial highways; we soon see them darting above our heads with their unmistakable flight. It looks erratic, but in truth it's the opposite as they chase down their prey, largely invisible to our eyes, with extreme precision. They are remarkably small – small enough to fit into a matchbox. My bat detector is really going for it now. The clicking sound suddenly becomes tinny. 'Soprano pipistrelle!' Alison exclaims excitedly as it pops up on her screen.

The bats seem entirely unfazed by our presence, which isn't surprising, Alison explains, because they have evolved alongside us. Like mice, pipistrelles are commensal and now largely depend on another species – humans – for their roosting and nesting habitats. 'Some species you'll only see by a waterbody, others in woodlands, but these are the species that live around us,' Alison explains; 'these are truly urban bats.'

Standing in this dark alleyway, which the police themselves seem to fear, Alison's primary concern is to avoid alarming anyone with our presence. I find this particularly touching, coming from a short and slight woman in her more senior years, armed only with a bat detector. But it's also this very darkness that makes the twittens so important for the bats. As nocturnal creatures, bats will only be triggered to leave their roosts when the light diminishes to a certain level. Pipistrelles can tolerate more light than most bats – up to about 14 lux, Alison notes precisely (by way of comparison, a family living room at night with the lights on is around 50 lux) – hence why they outcompete other species in urban areas. All the same, light pollution, or skyglow, remains an issue: the same issue that causes urban songbirds to stay up all night singing their hearts out until their energy is spent, or insects to orbit a cone of light until they expire, never to complete their lifecycle. 'The old orange streetlights were better for bats than the bright white light we have now,' Alison tells me. We demand a 24/7 city, and people understandably want to feel safe when walking the streets. But as the night becomes lighter and brighter, nature suffers.

Next, Alison wants to show me another part of the estate on the eastern boundary. On our way we pass more gardens, clearly well tended by the local residents. Alison points to a huge willow, well over twice as tall as the house behind it, with a mop of hanging branches. Soon it will be felled. Where some trees have already been cut the stumps remain and Alison has counted ten species of fungi. 'The fungi attract fungus gnats, which lay their eggs and beget more fungus gnats, providing a food source for the bats in the latter part of the year,' she observes.

We head through a walkway, in between rows of low-rise flats, and all of a sudden we're in a well-lit open space surrounding one of the huge tower blocks. There's an expanse of short-cropped grass, a concrete path, a bench and some nondescript flower beds filled with ornamental plants.

Our bat detectors fall silent.

This area – too well lit and too close to the main road for the bats – is one of the few 'green spaces' the developers are planning to retain once the estate has been demolished. Alison has brought me here to show me what an absence of life – a biodiversity desert – looks like. 'This is what the new estate will be,' she says, shaking her head sadly, 'just a sanitised space, with manicured hedgerows and gated courtyards. It's warehousing for people, with no habitat for humans or animals. There'll be nowhere for insects to feed, no place for a bird to nest or a bat to roost. We will be left with nothing.'

Alison has seen this all before – estates like this one, which she's surveyed in her work as an ecological consultant over the years. From time to time she's gone back of her own accord and found no signs of wildlife left. 'I'm old, you see,' she tells me, 'and one of the things about getting older is that you remember what the baseline for nature was decades ago. And you can see how it's shifted.'

This is all deeply political. Alison believes that it's not only the bats that will lose out; a community will be lost, gardens will be concreted over, humans crammed together like cattle. People with children at the local schools and relatives buried in the cemetery will be forced out. But there is another side to this. Sophie Kafeero, a resident and campaigner whose son was stabbed to death on the estate, was quoted in the local press in 2019. 'Living on a council estate, you're a human being,' she said. 'You've got needs. You want the rubbish to be collected, you don't want people selling drugs on the estate, you don't want people to be knifed on the estate, you want the buildings to be painted. You want the kids to have somewhere to go to play.' People like Sophie have long had enough.

It doesn't have to come down to a binary choice between humans and wildlife, Alison argues. There are a number of actions developers can take to mitigate habitat loss. These include dedicating more space to landscaping, a relaxed mowing regime, planting native trees and plants, providing bat and bird boxes and following specifications for more wildlife-friendly street lighting. While pipistrelle populations declined by an estimated 70 per cent between 1978 and 1993, they are finally showing signs of recovery, quite possibly thanks to the implementation of just these kinds of measures.

Before we leave, Alison introduces me to a resident named Damon, who has lived on the estate with his family for fifteen years. He seems undecided about the future of the estate, but Alison has certainly had an impact on him. 'If you keep your eyes open,' he tells me, 'there's lots of stuff here to see. Alison merely told me when to open my eyes.'

A few days later, while watering my front garden at dusk, I see pipistrelles out in my own estate, darting through the disused alleyways and into our close, where they whoosh past the flickering streetlamp and hoover up the haze of insects that surround it. I point them out to a group of our neighbours' children who are out playing: 'Look, bats!' As I'm about to head indoors I feel a strange vibration in my ear drum, and then I realise what it is – I can hear them calling.

RECLAIMING ROOFS

This chapter is about high places – the upper rather than the outer limits of our urban centres; the brick-and-mortar hills, concrete cliffs, steel-framed mountains and branchless trees, strung together with copper cabling. There are treetops in the traditional sense; sometimes even cliff faces too, where fragments of habitat remain amidst the seas of tarmac. And, of course, there is the sky itself, though we often forget to raise our eyes and look up.

The urban world has been designed almost exclusively by humans, for humans at every level. But things are starting to change.

In London, the change began with a small but determined gang of people, who spoke about the value of urban wildlife when few were listening, before I was even born. While many large organisations were focusing their efforts on the 'countryside' alone, in 1981 a group of such people established the London Wildlife Trust, giving official recognition to nature in the capital and fighting for it to be encouraged and supported.

A few years later, a man by the unforgettable name of Dusty Gedge got a job working on the regeneration of Deptford Creek in London's south-east. Dusty was a keen birdwatcher and an expert on black redstarts – one of the few birds almost exclusively breeding in cities that was afforded full legal protection at the time. He had originally moved to nearby Blackheath to join the circus, and found work running workshops for truants, but it just so happened that the funder of this project was also part of the Deptford Creek Regeneration Board. Knowing that he was a serious birdwatcher, the funder persuaded Dusty to join the board's ecology team as a bird recorder. As a result of his work in Deptford, Dusty quickly found himself on a mission to take hitherto unused and forgotten urban spaces and reclaim them for nature. He became a pioneer of the green roof renaissance.

I arrive forty-five minutes late at Dusty's house in Blackheath, having made the regrettable decision to drive instead of taking the train from St Pancras. I apologise repeatedly, but Dusty seems to have had little idea what time he was meant to be expecting me anyway. We had planned to head to a nearby IKEA, around ten minutes by car, which has a large green roof Dusty wants to show me, but he invites me in for a hot drink first.

As we head up the path that leads to his house, Dusty insists that I 'tread on some seeds' on my way in, pointing to a sheet on the ground as if there's nothing unusual about this request. Rather confused by his initiation ritual, I stomp my feet a few times on the sheet, which is filled with seeds, before following him into the house.

'You'll have to excuse the mess – I live on my own,' Dusty warns. The house is as eccentric as he is – he refers to it as an 'upside-down house' – the bathroom is on the first floor with the kitchen and living room above it. Each wall is painted in a different primary colour, and the kitchen is filled with unwashed crockery and ashtrays. A large papier-mâché bird's head hangs on one wall and a mural of Salvador Dalí adorns another. Dusty has been living here since the 1980s; 'It was falling down then and it's falling down now!' he jokes. He's had various flatmates over the years, he tells me, but they've all since moved out and he's the only one left. Rent controls mean he can afford to stay, despite the ongoing gentrification.

Dusty asks if I'd prefer tea or coffee, and I choose tea. He potters around for a while, heats up a pot on the hob and hands me a very strong coffee. I sip it slowly, not wanting to be rude, since he's gone to the trouble of making it.

Dusty offers to show me more of his seed collection, so we head out to the landing. All the way up the stairs are small plastic tubs and yoghurt pots filled with dried plant matter. In his bedroom are yet more seeds, stored in saucepans, Tupperware of various shapes and sizes, colanders and an ice box. 'Yellow-rattle, salvia, onion, chives, cow parsley, goat's-beard, vetches, honesty . . .' Dusty reels off the names of different plants; 'Take some honesty – it's good for your garden, especially if you want to attract orange-tip butterflies!'

Wherever he goes, Dusty picks up seeds. He begins by collecting cowslips in the summer heat and ends with marjoram in November as the days grow cold. He spreads them across his green roofs or gifts them to the people he meets to plant in their gardens. Most of his seeds come from east London – Blackheath, the Olympic Park and the Isle of Dogs – 'I always have little polythene bags in my pocket,' he tells me. 'My partner fucking hates it – seeds everywhere!'

Seed-collecting is perfectly legal in most instances. While uprooting entire plants without permission *is* against the law, according to the Wildlife and Countryside Act 1981, taking a few seeds is fine if it's for personal, non-commercial use and you steer clear of protected species such as rare orchids. You're not technically supposed to collect on most natural reserves, but Dusty says he's never been challenged. 'What you do is you don't take everything,' he advises, 'so I'll pull up an oxeye daisy, and I'll let a load of the seeds go as well as keeping a few for myself.'

On the windowsill are yet more seeds, which are drying out in the summer sun, next to which sits a small box designed to provide a home for passing solitary bees. '*Osmia cornuta*!,' Dusty announces, having seen that the box has caught my eye; 'they're nesting in there right now.'

Osmia cornuta, as it turns out, is the Latin name for the orchard mason bee, one of the earliest bees to come out of hibernation, usually in February. They're pretty-looking little creatures, 10–15 centimetres in length with an orangey-yellow abdomen and a black thorax and head. The females have horns and the males a long pair of antennae and a patch of white fuzz on their faces resembling an overgrown moustache.

The species is new to the UK, officially recorded for the first time in Barnes, south-west London, in 2016. But it was spotted before that, in 2014, when one emerged from this very bee box. Dusty emailed a pre-eminent bee expert at the Natural History Museum with a photograph, which he misidentified. Two years later the bee expert realised his error, but it was too late by then and Dusty never got the credit for his discovery.

Nevertheless, his bees are still attracting attention – sometimes he spots entomologists walking purposefully up the street, eager to catch a glimpse of the rare polli-nator. 'I shout, can I help you? I know you're looking for bees! They're nesting in here. Come up and have a look,' Dusty laughs.

Before we head back downstairs Dusty picks up one more pot of seeds to show me. These ones are special, he says – they're from his favourite plant, hare's-foot clover, a delicate wildflower that thrives in dry grasslands, coastal areas and, as it happens, on green roofs. Its pale pink flowers are surrounded by downy hairs, giving the appearance of a soft paw. It's pretty much impossible to buy hare's-foot clover commercially so Dusty's seeds are always in demand. After collecting from a single spot over eighteen years ago, he's managed to successfully sow it on rooftops across the capital. 'I call it spreading the love,' he declares. (I'm relieved he didn't opt for 'sowing my seeds'.)

This finally brings us on to the topic that led me here in the first place – green roofs. But to tell the story of Britain's green roofs, we must also turn to the story of black redstarts – because through Dusty's work the two have become inextricably linked.

Though more numerous in Continental Europe, in the first half of the twentieth century black redstarts were extremely rare in the UK, with just a handful of breeding pairs recorded, including one couple that notably bred at the Wembley Exhibition Centre in 1926. But towards the end of the Second World War this small, robin-sized member of the thrush family really took to the city. Black redstarts began to trade cliffs, hillsides and mountain gorges for the bombed-out, abandoned buildings and stony wastelands left behind by the Blitz. They started to breed in London, other British cities including Sheffield, and across the Black Country, where the collapse in manufacturing left plenty of disused buildings to commandeer. By 1964 there were sixteen pairs nesting in Cheapside, and as areas of central London were cleared up and rebuilt the birds gradually spread out across the old docks of east

London, where they found empty warehouses and open water, with plenty of insects to eat.

But the black redstart's success looked like it might be relatively short-lived as these sites were earmarked for redevelopment in the 1970s and 1980s. And this is where Dusty's work at Deptford Creek comes into play.

Ecologists like those in Dusty's team had already begun to take note of the wildlife they were encountering in areas of brownfield and sites of vacancy and dereliction. The creek was just such an area, and Dusty quickly discovered three or four pairs of black redstarts. It would have been an offence to disturb them. So for the Deptford Creek regeneration work to go ahead, a plan for the black redstarts first had to be put in place.

'We had this idea in the ecology team – well, why don't we just shove it all up on the roof?', Dusty jokes, 'and the rest, as they say, is history.'

Dusty quickly discovered that because of their protected status, black redstarts were a useful lever that he could use to prevent developers from acting with impunity: he could mitigate some of the damage they were causing to local biodiversity by forcing them to replace lost habitats with new ones on the roofs of their buildings. 'In the 1990s I spent a lot of my time wandering around King's Cross in the early mornings when it was still full of sex workers and druggies, trying to find black redstarts,' Dusty recounts. 'I'd find them deliberately and get whole developments shut down as a result.'

The tale of the black redstarts as Dusty tells it is one

of renewal. When the redevelopment started in London after the war, the black redstarts moved out to the docks, which had been abandoned following the collapse of the empire. As regeneration began in the old docks and factories, they moved to the brownfield sites. And when the government decided that it was time to build on the brownfield sites, the green roofs were simply the next phase of renewal.

Green roofs benefit not just black redstarts, but a whole range of invertebrates and birds, providing giant stepping stones for aerial organisms to navigate the city. They also help to store rainwater after intense storms, where it might otherwise take hours to dissipate, causing all manner of problems in the meantime. As the impact of climate change worsens we're likely to experience greater numbers of severe weather events, so this is surely going to become increasingly important.

Climate change means we're also likely to experience increasing temperatures, which are exacerbated in cities by the urban heat island effect. Dark, impermeable surfaces like roads and pavements absorb solar radiation. Tall buildings form a wind block, as well as adding multiple surfaces that reflect or absorb sunlight. Vehicles, industrial activity and air-conditioning units generate waste heat. The London Plan, put together by the Mayor of London and officials at City Hall, predicts that by 2030 heatwaves in the capital will cost lives, so the Greater London Authority clearly sees this as a serious problem. Green space is known to mitigate the impacts of the urban heat island effect, and green roofs can play a big part in this; the plan now calls for new development proposals 'to include roof, wall and site planting, especially green roofs and walls where feasible'.

As a result of all these factors, Dusty, who is now President of the European Federation of Green Roofs, found himself at the forefront of a green roof revolution. In 2010, London had 200,000 square metres of green roof. Today the number stands at over 1.5 million; 'and I've mapped them all, so I should know', Dusty grins.

I'm keen to see a green roof for myself, so after about an hour of chatting Dusty and I finally jump in my car and head to IKEA. As we drive through Blackheath he shows me the 'bee roads' that he's planted across the green – strips of tall grass and wildflowers that are spared from the mowers' sharp blades, allowing pollinators and other invertebrates to commute in safety.

Dusty's sporadic directions and my slightly nervous driving aren't the best combination and we end up accidentally cruising down a bus lane, forcing me to take a slightly hairy u-turn, before we make it to the IKEA Greenwich car park. He hands me a high-vis jacket 'to make us look official' and tells the security guards we're here to 'inspect the roof'.

Despite the building only being a few storeys high the view is spectacular – we can see right across the city. The towers of London's financial district at Canary Wharf feel incredibly close and I'm relieved I brought my sunglasses – the glare is dazzling.

IKEA Greenwich is a good introduction to green roofs, Dusty explains, because it's split into different sections, demonstrating a number of different approaches. The first is what's known as a 'sedum blanket system', a particularly popular way of greening a roof because it's easy for developers to buy and install. It's pretty, too, forming a dense, uniform carpet of small, rubbery plants in reds, oranges and greens, which sprout delicate white flowers when in bloom. The sedum blanket here is a particularly good example, Dusty tells me, but from an ecological standpoint this method of planting isn't particularly valuable. It's a monoculture, with the sedums only flowering for about three to four weeks a year. A comparable area planted with wildflowers can bloom from March to October, supporting pollinators for a much longer period.

Because this roof has been designed with public use in mind IKEA has insisted that the parts that are most visible are kept tidy, but Dusty has persuaded them to add some areas of compacted sand to this section for solitary bees to nest in, as well as small piles of rubble just out of sight, from which various wildflowers are already starting to sprout. He points out a clump of evening primrose, covered in waxy, acid-yellow flowers, grown from seeds he collected outside a local school. While you might be surprised that anything much can grow from a pile of rubble, it's actually an ideal environment for wildflowers that flourish in rocky or chalky terrain.

The next section of the roof, Dusty tells me, is focused on health and well-being, which has gained increasing attention following the coronavirus pandemic and the periods of lockdown during which we've all realised just how precious our green spaces are. There are apple trees and strawberries, cabbages and red onions – the beginnings of a community food-growing initiative, connecting people with what they eat at source.

We don't spend too long in the third section, which is probably the area where most visitors to the roof spend their time. It's planted very deliberately with attractive-looking grasses and non-native plants, which only provide real value to humans; 'It's about having a garden,' Dusty explains, 'it is what it is.' To his chagrin, the ground has been AstroTurfed; 'They were worried kids might get their knees dirty,' he explains with an eye-roll. 'It's atrocious stuff – to clean it they put all these microplastic pebbles on it and then wash them out into the environment. And yet you'd be amazed how many people try to sell me AstroTurf on LinkedIn – drives me fucking mad!'

We move on to the final section, which is split into two. The first half was initially turfed with wildflowers, but all that remains are a few sad, scorched-looking clumps. 'I've managed to persuade them not to fucking irrigate it,' Dusty explains, 'because it's not necessary – over time these grasses will come back.'

To irrigate or not to irrigate is an area of much contention in the green roof space, and even more so when it comes to green walls. Understandably, the two get lumped together, but they are actually quite different. 'I don't

do green walls,' Dusty tells me with evident contempt
for the entire concept. Many of the lush-looking green
walls you can find across British cities are crammed full of
non-native plants. They're essentially an artifice – a green
illusion, where any benefit they may bring is wiped out
by the large volumes of water needed to maintain them.
'These green wall people – most of them think they've got
dicks bigger than anyone,' Dusty grumbles; 'I'd rather let
my green roofs die and then grow back again than waste
all that water.'

The second half of this section of roof is faring much
better. Instead of using turf, Dusty built little piles of
rubble, which he scattered with the seeds of hardier plants
that are better equipped to survive in times of drought.
Now all that remains is to let nature take its course. This
approach requires patience, but in five years' time, he tells
me, it will be spectacular.

We can already see little patches of viper's-bugloss,
with its spikes of conical-shaped flowers that turn from
pink to vivid blue. With a slight imaginative leap its
stem – stippled with spots and tiny hairs – resembles a
viper, hidden amongst the foliage; perhaps that's why
in ancient times its roots were used to treat snakebites.
There's chicory too in its wildest form; the woody stems
and pale-purple flowers bearing little resemblance to the
cultivated plant used in salads that's forced to grow underground to achieve its pale, crunchy leaves. In the past,
wild chicory roots were used as a coffee substitute and a
natural sweetener; its leaves add a bitter flavour to soups

and stews. Dusty collected the seeds from a roadside verge just off the A2, he tells me.

Then there's wild marjoram with its clusters of tiny pink flowers, toadflax, which looks like a miniature custard-coloured snapdragon, and quaking-grass, so named for the way in which its bug-shaped seed heads quiver in the wind. There's bird's-foot-trefoil with claw-shaped seed pods and flowers like yellow slippers, and of course there's Dusty's pride and joy – the hare's-foot clover.

Dusty completes our botanical tour with a little patch of kidney vetch, its low-growing furry cushions already sprouting clusters of yellow flowers.

Kidney vetch is of particular importance because it's the only source of food for Britain's smallest butterfly, the small blue. Small blues are declining in many areas and Dusty was met with disbelief by ecologists when he claimed to have caught sight of the rare species on one of his green roofs. Fortunately he managed to gather video evidence. 'Often ecologists look at a roof and think it's different. Because it's "not natural" they assume that wildlife won't like it up here. But what's the difference between this roof and the heath we just drove through?' he insists.

While the relentless drive of urban development is unquestionably responsible for habitat loss, Dusty's tale of green roofs and renewal is the perfect example of how we can start to turn the tide if we use our imagination and look up. Green roofs are characteristic of the boundary-blurring that takes place in urban environments, between what is human-engineered and what is 'natural'. Dusty leaves his roofs to grow and evolve with as little management as possible. The seeds he sows have been carefully collected, not bought, thriving amidst piles of rubble that mimic the habitats he hopes to recreate. Green roofs might not appeal to everyone's concept of what it means to be wild, but this is of little consequence to the black redstarts and the small blues. In a world where precious resources are increasingly hard to come by, they're happy to make use of these patches of green in the sky.

CHAPTER 3

Look Out: Island Life

I never dreamed that islands, about fifty or sixty miles apart, and most of them in sight of each other, formed of precisely the same rocks, placed under a quite similar climate, rising to a nearly equal height, would have been differently tenanted.

Charles Darwin, *The Voyage of the Beagle*

'No man is an island', wrote the seventeenth-century poet John Donne; 'Every man is a piece of the continent / a part of the main', his anti-isolationist words taking on a new contemporary relevance as our own little island fights to find its place in the world. And yet, however internationalist we may feel in outlook, most of us have at some point longed to escape to an island of our own, pristine and untouched, cocooned by a moat of protective waves. We've all sought a place of isolation; a retreat from the constant noise of social media, politics, digital advertising, influencers, rising inequality, pandemics, drought, climate change.

Ask most people to shut their eyes and picture themselves on such an island and they'll describe somewhere distant and exotic, the kind of destination marketed to honeymooners – palm trees, white sandy beaches, crystal waters and coral reefs: Mauritius, the Seychelles, Bora Bora, the Maldives. (Take away the five-star resort and the reality isn't always what it's cracked up to be – I remember reading an article by a contestant on the Channel 4 reality TV show *Shipwrecked*, reflecting on his experience. The title gave a pretty good summary: 'Crotch rot and mountains of maggots: my five months of hell'.)

As a child, my parents, who certainly didn't have the means to take us to Bora Bora, planned a trip to the Isles of Scilly, 28 miles off England's Cornish coast. It was the first time I'd ever travelled by air. We gazed down at the fragments of land surrounded by blue-green water, my mother frantically reaching for her camera to take a picture through the window. While there, we took a boat trip to Samson, the largest uninhabited island in the archipelago. It has no quay, so visitors are forced to disembark by walking the plank. And we visited the Beady Pool on the eastern side of St Agnes, named after the tiny earthenware beads from the wreck of a seventeenth-century Dutch cargo ship that are still occasionally found there.

That trip to the Scillies, along with countless hours spent watching nature documentaries, sparked an unshakeable desire in me to visit islands of increasing remoteness. Borneo, home to orang-utans, but also to the bizarre-looking proboscis monkeys, with what must be the strangest noses I've ever seen on a living creature. Magnetic Island, named after the 'magnetic' effect it supposedly had on Captain Cook's ship compass as he travelled up Australia's east coast in 1770. The Andaman Islands, nestled in between the Bay of Bengal and the Andaman Sea, known for their colourful coral reefs. And, most remote of all, Easter Island and the Galápagos, moulded by the mighty volcanoes of the South Pacific.

When the captain of Charles Darwin's ship, HMS *Beagle*, first set foot on Galápagos in September 1835 he described 'a shore fit for Pandemonium'; to his eyes the strange black shapes, formed over several millennia from molten lava, took on a hellish quality. But for Darwin those islands held the key to unlocking the mysteries of evolution, and it was their remoteness that aided his discovery. When studying four species of Galápagos mockingbird – similar to those found on mainland South America – Darwin came to the realisation that species change and adapt over time in response to the habitat in which they find themselves.

It was not the mockingbirds, though, that were to end up the heroes of this evolutionary tale, but a smaller group of birds, which became known as Darwin's finches. Darwin was unable to make much of them at the time, other than to observe their vast array of beaks. On his return he presented thirty-one specimens to the ornithologist John Gould, who identified thirteen species, but beyond that the riddle of the finches and their varied beaks remained unsolved. It wasn't until 1947 that an ornithologist named David Lack made the connection between the birds' beaks and their highly specialised feeding habits – from the large ground finch, which uses its short, chunky bill to crack nuts, to the blood-sucking vampire finch, with a razor-sharp beak that can pierce through the skin of unfortunate Nazca and blue-footed boobies, the islands' enigmatic seabirds.

The Galápagos Islands continue to reveal evolutionary secrets, with the finch populations in particular transforming at speeds which Darwin would never have believed possible. A decades-long study of the birds by British ornithologists Peter and Rosemary Grant has shown that over the course of just twelve months the beaks of a single species can change dramatically based on such factors as food availability (some foods favouring smaller, more delicate beaks; others requiring a longer beak to extract nutritional value), competition and hybridisation between species.

But you don't need to travel to the far-flung corners of the earth to visit an island, or to discover the impact (evolutionary or otherwise) that island life can have on its wild inhabitants. Cities are in many respects islands in and of themselves, with their own microclimates, thanks to the urban heat island effect. This has been known to impact our urban wildlife in the most fundamental of ways, bringing biological clocks forward, with some plants flowering earlier in the city and birds and mammals mating and rearing their young weeks ahead of their country-dwelling counterparts. Some urban species even manage to raise additional broods as a result.

Towns and cities are also full of smaller islands – parks and green spaces, where certain populations can become completely cut off from the rest of their species, the effect of which we're only just beginning to understand. Meanwhile, residential gardens provide vital landing strips for airborne creatures to rest and refuel. Even the smallest scrap of earth – a tree pit (the hard ground around the base of trees) enclosed by a sea of tarmac – or a crack in the pavement can provide value to a range of different species. As relatively new ecosystems in which nature is carving out its niche, our urban islands are living experiments ready to be explored, with results which Darwin himself could never have anticipated.

SPONTANEOUS PLANTS

In spring 2020, when the UK was in the midst of its first coronavirus lockdown, graffiti began appearing in chalk on the pavements of east London suburbs. Technically, it's illegal to use chalk on public paths or highways without permission (tell that to the lawless children scrawling flowers and rainbows across my estate), so for a time the chalkers remained anonymous. But their work soon went viral. 'To whomever is chalking names and descriptions of trees on the pavements across Walthamstow. I love you,' tweeted Walthamstow resident Elizabeth Archer to a chorus of 150,000 likes and retweets; 'This made my heart sing today.'

The mystery Walthamstow chalker turned out to be self-employed forest school practitioner and mother of five Rachel Summer, who was hoping to bring a little light to her neighbours' daily walks during lockdown. 'Silver birch – provides food and a home for over 300 insect species', she wrote in big pink letters. 'Sycamore – a real survivor. Grows ANYWHERE!', she added, with an arrow pointing towards a metal fence, through which a flurry of green leaves could be seen escaping. 'London plane (my favourite!) Takes pollution out of the air', she annotated at the base of a trunk.

Rachel was inspired by a growing movement of international rebel botanists, self-appointed cheerleaders for the neglected, forgotten and downtrodden plants of our urban landscapes. These tiny islands of green can be found colonising cracks in pavements, sprouting from tree pits, escaping from gaps in walls and clinging to brickwork. The chalking was an idea that began in France, where the use of pesticides was banned in streets, parks and public spaces in 2017, and private gardens in 2019, allowing the incidental plants that pop up in urban environments the chance to flourish.

Meanwhile in east London's Hackney, French botanist Sophie Leguil was also putting chalk to pavement as part of her 'More than Weeds' campaign. In 2018 Hackney became the first local authority in the country to trial 'glyphosate-free zones', dropping the use of the harmful weedkiller. Along with the damage it causes to biodiversity, glyphosate has been found to be 'probably carcinogenic' to humans by the International Agency for Research on Cancer. At the same time as banning glyphosate, Hackney sanctioned chalking – for Sophie's purposes at least – making it the perfect place to launch her campaign.

It's a drizzly day in early summer when I call Sophie

on FaceTime. After a long stretch of good weather it's a sudden reminder that the sunshine is far from guaranteed. Sophie is back in France with her parents for a few months so I've arranged to take her on a virtual tour of my surroundings to see if we can find some interesting 'weeds'.

I don't know what I expected a botanist to look like – old, bespectacled, and tweed-jacket-wearing, perhaps – but Sophie is young and smiley, with dark, wavy hair that cascades over her shoulders. I like her straight away – she's easy to talk to, chatty, speaks very fast. I start by asking what makes a plant a 'weed' – a question that has perplexed me since we moved to a house with a small garden a couple of years ago and we came to the realisation that something would have to be done to maintain it. The garden was already relatively mature, so I've been afraid of weeding for fear of taking out the wrong thing (or that's my excuse at any rate). Sophie immediately reassures me here; 'A "weed" is just a plant that's growing out of place.' So I now have permission to leave the weeding for another year.

But this becomes a little more complex in public spaces; who defines what is or isn't out of place in our parks, cemeteries and commons; on roundabouts, in tree pits and between the gaps in the pavement? Most councils, it seems, consider anything that they haven't planted themselves to be a weed. But a lot of self-seeded plants are wildflowers – the sort that people put considerable effort into encouraging on green roofs and carefully maintained meadows. 'If you were in the countryside you wouldn't call them weeds,' Sophie observes, 'so why are they suddenly out of place in an urban environment?'

Sophie opts to replace the term 'weeds' with 'spontaneous plants'. Leaving 'spontaneous plants' where they grow results in many benefits, she argues, above and beyond the fact that they brighten up the urban landscape. They provide tiny island habitats for insects; food for caterpillars from early spring onwards and a source of nectar and pollen when they begin to flower. Then, when the flowers are over, birds feed on their seeds. They create oxygen and they have the power to filter out pollution, making it counter-intuitive to remove them at one stroke while implementing clean air legislation at another. Plantain, one of the most common 'weeds', with its rosette of broad green leaves and poker-shaped seed heads, is especially good at capturing pollutants.

There are some plants that need to be removed, of course, if they're obstructing a path or creating a potential hazard for people with disabilities. But simply allowing plants to populate tree pits alone would vastly increase the amount of urban green space. 'I did a bit of research and found that London has 900,000 tree pits,' Sophie tells me. 'If plants were allowed to grow in each of those 60-by-60-centimetre pits, it would equate to the same size as St James's Park!'

To get to such a place a cultural shift is required. That shift is already taking place in France, where, following the glyphosate ban, the local authorities have been on a communications drive to change the public's perception of 'weeds'. The French are getting used to seeing more plants growing freely in urban environments with an increasing understanding that this is a deliberate choice and not wilful neglect or a sign that their council tax payments are being misused.

Meanwhile, here in the UK it was the coronavirus pandemic that briefly gave our urban plants the chance to flourish. During the first lockdown, many of the contractors that perform the general disservice of spraying all signs of life with chemicals and wage war with strimmers and mowers were told to stay at home. Ordinarily this process would begin in spring, with the plants removed just before they have the chance to flower. But with the lockdown lasting from March until early summer the pavements suddenly sprang to life.

Sophie and I begin our tour in one of my disorderly flower beds, where new plants seem to pop up by magic each spring. I focus my phone's camera on a couple of rather leggy-looking plants with small, bell-shaped flowers like old-fashioned lampshades. One plant has pink lampshades, the other purple. I've noticed that the buff-tailed bumblebees that have been nesting in a hole at the bottom of our fence are particularly enamoured with them. Some of the bees collect pollen in the conventional way, stuffing themselves inside each flower. But others have developed a sneaky shortcut to extract the sweet nectar by biting a tiny hole at the top of each bell and sipping it out.

'Aquilegia!' Sophie exclaims, before revealing the plant's English name, columbine, 'or granny's bonnet because of

the shape of the flowers'. To modern sensibilities columbine is a bit 'weedy', Sophie concedes, but it was loved
by the Victorians and technically remains a garden plant.
Columbine produces a large number of seeds, as I learn
for myself a few weeks later, so it spreads easily and can
be found in paving cracks and other unexpected places.
Its blooms are various shades of pink, purple or white,
but it often hybridises – you can start off with purple and
white one year and end up with pink the next.

I move us on to what I'm confidently able to identify
as a poppy, although it has recently shed its trademark
red petals. 'But this isn't your typical cornfield poppy,'
explains Sophie, 'it's an opium poppy.' Opium poppies are
one of the most common urban plants due to their har-
diness – originating from Asia, they can survive through
periods of extreme heat and drought – and their tiny seeds
are easily carried by the wind, or spread by birds from
one urban island to another. These are the same seeds that
are used in baking, hence why it's also sometimes known
by the much more wholesome name of 'the breadseed
poppy'.

Next I show Sophie a plant I first noticed last year. I
watched as it grew up and up and up, waiting to see what
it might look like in bloom. Eventually it threw out a
careless little cluster of yellow flowers at the top, like one
of those fireworks you buy in the supermarket that raises
expectations, only to culminate in a tiny puff of coloured
smoke. Following this disappointment, I tried to rip it
out and ended up with even more of it for my trouble.

'That's fleabane,' Sophie tells me; 'it's a North American weed, but very common in London. You find it a lot in patches along railways – in dry, rocky places. It can grow really tall, so it's impressive in size, but the flowers aren't up to much.'

I move along to the bed that runs along our back fence, which is dominated by a large, wayward rose bush. On my hands and knees, I locate a tiny plant beneath the bush with purply-red heart-shaped leaves that creep along the ground. Sophie identifies it immediately as oxalis, another common plant that springs up in urban settings. 'Have you seen its flowers?', she asks; 'they're very striking – a vivid yellow that stands out against its deep-red foliage.' The origins of oxalis are unclear; it may have come from Asia, but its ability to grow in tough conditions means that it can now be found across five continents – I later find another clump growing in between the paving slabs on our driveway.

Garden tour complete, we head out of the gate and into the estate of near-identical red-brick houses, Sophie's face still looking out from my phone. My neighbours are fairly used to my antics by now, but even so I must look a little odd as I crouch down, pointing my phone towards cracks in the paving. I find a small plant growing flat to the ground in between the slabs in a parking bay. Its leaves are green and feathery and it has yellow, dome-shaped flower heads, like daisies without the petals. 'That's pineapple weed,' Sophie pipes up, directing me to crush one of the flowers between thumb and forefinger and hold it under my nose – it smells remarkably like pineapple. Otherwise known as wild chamomile, pineapple weed is another plant that thrives in tough environments, bursting out of concrete, rubble and areas of compacted soil. Its leaves are edible and can apparently be added to salads and its flowers used to make herbal tea, but my dog-owner instinct tells me not to trial any recipes using this particular patch.

Along the same stretch of paving I find a taller plant with slightly pointed green leaves which jogs something in my memory. It looks like something I used to pick as a child to feed our giant rabbits, with tiny yellow cylindrical flowers. 'Yes exactly!' Sophie replies; 'that's groundsel. It's very popular with rabbits, birds, lots of animals in fact.' In the nineteenth century there was a living to be made from collecting weeds such as groundsel from wealthy people's gardens, which could then be bundled up and sold in the street as food for the caged songbirds that were so popular with the Victorians. After our call, Sophie sends me a link to an old engraving entitled *The Groundsel Man*. In it a man in a top hat with unkempt hair and whiskers grasps a bunch of weeds, with more spilling out of a basket slung over his shoulder. 'Chick-weed and Grun-sell!', the caption reads.

In this one area of just a few metres in length I continue to find a good deal more pineapple weed, a tiny island of redshank (not to be confused with the red-legged wading bird) with long, tapered leaves and pink flowers crowded onto little spikes, and a clump of willowherb, its bright pink four-petalled blooms standing in contrast with the concrete backdrop. Known as 'fireweed' in its native North America, where it's the first plant to colonise after a forest fire, it became known in Britain as bombweed, due to the speed at which it took over abandoned bomb sites after the Second World War.

I also encounter a patch of common knotgrass (not in fact a grass), with long, semi-erect stems and tiny pink and white flowers; its Latin name, *Polygonum aviculare*, refers to the fondness of small birds for its seeds.

Moving on from the parking bays, I head towards the canal at the end of the close, past a well-kept communal flower bed that my neighbour Georges has been tending (recently, while doing some weeding he found a stash of a rather different kind of weed hidden amongst the foliage.) On the stone wall that separates our estate from the canal below I find a large patch of hardened brown moss which, according to Sophie, is surprisingly good at filtering out pollutants from the air. A small plant with red-tinged leaves sprouts from an area where the wall is crumbling. Each pretty pink flower has five petals, with white stripes radiating from the centre. I pause to admire it. 'Herb-Robert, my favourite!' Sophie interjects. 'It's probably the most common urban plant.'

In traditional herbal medicine herb-Robert had uses as wide-ranging as a treatment for nosebleeds and stomach upsets to an antiseptic and insect repellent. It has an extended flowering period from May to September, Sophie explains, making it especially useful for pollinators. The name is thought to originate from an association with a house goblin of English folklore, Robin Goodfellow, also known as Puck, who appears in Shakespeare's *A Midsummer Night's Dream*. Herb-Robert has a particularly unpleasant smell and Puck is a renowned mischief-maker – perhaps the two are connected.

As our little tour concludes I realise how infrequently I stop to take in the detail of my surroundings and how little I know about the species of plants I encounter on a daily basis. Sophie tells me that, since launching her More than Weeds campaign, she's been flooded with messages from people whose perceptions she's changed, their eyes now opened to the diversity of life within our urban landscapes. Our cities are a colourful melting pot of cultures, communities and ideas and this includes our urban plants – some have evolved here, some are immigrants that have been deliberately introduced, others have hitched a ride on a piece of luggage, embedded in fur or feathers or attached to the sole of a shoe.

After my call with Sophie, I make a conscious effort to look out for these tiny islands of greenery in the most unlikely places. I download an app called Picture This, which allows me to take a snap of a plant, which it then miraculously identifies. It's a bit like having a botanist on my phone at all times (although it's not, of course, as engaging as Sophie). I start spotting spontaneous plants on my morning dog walk, at the bus stop, on my way to the Tube, in supermarket car parks, by the side of the towpath, dotted across housing estates and children's play parks; sprouting from cracks, crevices and gaps in pavements and brickwork. In fact, there are few places where plant life can't be found, once you start looking.

NEEDLES IN A HAYSTACK

It's ten o'clock on a clear July evening in London. Despite the ever-present glow of the city, I can actually make out some stars.

My dad and I are standing on the outer circle of Regent's Park as we wait for Mark Rowe, Assistant Park Manager, and Bryony Cross, Volunteer and Programmes Manager for the Royal Parks. The temperature is warm but not stifling, with enough of a breeze to necessitate wearing a jumper under my denim jacket. For once I have dressed correctly. We watch as the last few people leave the park, beers and picnic blankets in hand.

We're not waiting for long before Bryony and Mark arrive. Bryony, who is new in the role, is the younger of the two, chatty and enthusiastic. Mark is quieter; he's an old hand at park management, having worked for Islington Council for some time before taking on his current role five and a half years ago. He passes round some torches while Bryony takes out a hand-held infrared camera, which we hope will help us to locate a small and often elusive spiny creature that resides in the 395 acres of open space that make up Regent's Park.

Hedgehogs can be found in the outer, more suburban parts of London that form a halo around the city – on Hampstead Heath and in boroughs such as Redbridge and Bromley – but the hedgehogs in Regent's Park are the last remaining population in central London. According to a little book called *Wild Life in the Royal Parks* by the ornithologist Eric Simms, hedgehogs were present in Hyde Park, Kensington Gardens and St James's Park in 1974, and Dr Nigel Reeve, former Head of Ecology at the Royal Parks, says park managers reported seeing hedgehogs in Hyde Park into the 1980s. But no one is quite sure when those populations disappeared for good.

Why they disappeared is perhaps an easier question to answer. Over the past decade or so, ecologists have begun to realise just how important connectivity is for Britain's hedgehogs. This has led to a drive for both individuals and property developers to create tunnels and gaps in fences, known as 'hedgehog highways', allowing these small mammals to travel between gardens and other green spaces. But for many urban populations, including the Regent's Park hedgehogs, this is impossible. They are marooned, surrounded on all sides by impassable obstacles. Travelling beyond their island confines to the nearest green space – leafy Primrose Hill – would require a perilous journey across busy roads and the Regent's Canal. The hedgehogs would have to find their way to the north-east corner of the park, where there is a bridge, and brave around half a kilometre of impermeable concrete and tarmac – a nigh-impossible feat for a creature weighing just over a kilo, whose only form of defence is to curl up tightly into a ball. When it comes to facing off a two-ton metal predator roaring past, wheels spinning, horn blasting, hedgehogs simply don't stand a chance.

I ask Nigel whether anyone has thought about building a green bridge between the two parks. 'Thought about it? Oh yes,' he replies, 'but taken the practical steps? No. It would be outrageously difficult to do.'

Being isolated left central London's island-dwelling hedgehogs vulnerable to chance events, like the 1976 drought, which Nigel expects led to high mortality. Normally, when scientists look at a given population they find that it's made up of a number of sub-populations, because when one group suffers the area is re-colonised by other nearby groups and on average the numbers remain stable. But that can't happen in fragmented urban environments. 'All you need is some bad luck and the population loses its viability and dwindles out. That's almost certainly what's happened in London's central parks,' Nigel tells me.

Genetic testing has revealed a restricted gene pool amongst the Regent's Park hedgehogs, suggesting that they belong to an original population that has been sequestered in the park for a very long time. This poses further questions about how best to protect them. 'There's a general ecological understanding that most people subscribe to,' Nigel explains, 'that if a gene pool is limited then the population is more vulnerable, because it's got less variation to adapt to changing circumstances or new diseases. But in this population that's not necessarily true.' He gives the example of hedgehog populations that have been introduced to Scottish islands and have done extremely well, despite being founded by a very small number of animals. In fact, introducing new individuals now to bring in 'fresh blood' could have a detrimental effect, with the risk of introducing new bacteria, viruses and parasites at the same time. The new hogs might outcompete the original ones, or they might suffer the same fate if things start to go downhill. There's even a chance that the existing bloodline have adapted to their specific environment in a way that could be diluted or lost if they were to interbreed with outsiders.

So the Regent's Park hedgehog population is both unique and fragile. It's subject to a decade-long study (getting funding for a research project of that length is no mean feat), which includes annual hedgehog counts supported by dozens of willing volunteers – there's always a long waiting list, Nigel tells me. Hence why Bryony and

Mark are so well kitted out and accustomed to night-time excursions into the park, long after the Tannoys have announced that it's closing and the gates have been locked to the general public. (Of course, a few people find their way into the park at night for more nefarious, non-hedgehog-related reasons. When I mention my planned excursion to a friend he confesses that, having snuck in after hours with his girlfriend, they were startled by the sound of something snuffling around their picnic blanket. His girlfriend was horrified, assuming it was a rat, but according to my pal it was distinctly prickly.)

We only make it a few metres into the park before Mark has to run back with his keys to let out the last remaining stragglers, but then it's just us. I wasn't quite prepared for how different – how bewitching, how captivating – the park would feel at night. While there's plenty of traffic noise in the distance, it's remarkably still. It's a place I know well, but under the cover of darkness it's rendered unrecognisable, shadowy and mysterious; trees seem to sneak up on you and you have to watch where you step to avoid tripping on a tree root, or startling a hedgehog.

Hedgehogs tend to nest in the undergrowth, sometimes using the nest boxes that the team have installed for them. 'As well as the counts, we do nest box checks too, during the day,' Bryony tells me. 'We found a family of hoglets in one last year. You definitely get some strange looks, though, crawling out of hedges on your hands and knees.' But tonight we're looking for adult animals, so our best bet is to check the grassy edges where they often venture out to feed. Bryony shows me how to use the infrared camera, which picks up the warmth of bodies in the dark from up to 20 metres away. She warns me not to get too excited because most of the time the little blobs of red that appear on the screen are not hedgehogs but coke cans that have retained the heat of a sunny day. Concrete, and even tree roots, can have the same deceptive effect.

It's important to use your ears too, Bryony notes ('Which is difficult if you're talking all the time,' Mark teases) – 'Listen out for something bigger than a bird making a kind of shuffling sound,' she directs. Without an infrared camera it's still perfectly possible to find hedgehogs by keeping your ears pricked and sweeping a torch across the ground in front of you as you walk – in fact, this is the method used by many of the hedgehog count volunteers, as there are rarely enough infrared cameras to go round.

Over the years the team have identified various 'hotspots', where the hedgehogs like to forage. Like many urban-dwelling species, hedgehogs are generalists and will eat whatever they find lying around, but the majority of their diet is made up of invertebrates: most commonly worms, slugs, beetles, caterpillars, earwigs and millipedes. In 2017 Bryony ran an invertebrate survey in the park to see how much food was available. 'We wanted to get people to join the dots between the presence of invertebrates and the success of the hedgehogs,' she explains, 'but getting them to look for slugs and snails was a much harder sell!' Fortunately, once Bryony had corralled enough volunteers she found that from a dietary perspective at least the Regent's Park hedgehogs are pretty well catered for. 'None of this would happen if it wasn't for the volunteers,' Bryony adds; 'the power of citizen science is just immense.'

The team has met with a few surprises along the way too. One year, Bryony tells me, they found a hedgehog nest inside the anteater pen in London Zoo, in the midst of some rushy vegetation. And during the first hedgehog count in 2014, while they did a fairly broad sweep of the park they missed out the zoo car park in the north-east corner. One day a volunteer who also worked at the zoo asked why the car park wasn't included in the survey, following a tip-off from the night-watch staff. 'We said we simply didn't think we'd find anything there,' Mark admits. As it turns out, the car park was one of the hedge-hogs' favourite haunts, surrounded by their preferred kind of habitat, with low-lying brambles, patches of grassland and plenty of trees to provide leaf litter to hide amongst.

So for a while the car park became a key focus until, sadly, a large chunk of it was repurposed as a holding area for the vast lorries that now thunder up and down between Regent's Park and Euston Station, where the new HS2 high-speed rail link is being built. Mark and the team made their protestations. 'But the thing is,' Mark tells me, 'they were already ploughing through ancient woodland and triple SIs [Sites of Special Scientific Interest] and we wanted them to change their plans because of some hedgehogs in a car park?' With additional local pressures, HS2 were persuaded to build the first ever civil-engineering hedgehog tunnel, so there was a win in there somewhere, although it remains to be seen whether the hedgehogs will use it.

After around half an hour of searching, the infrared camera has picked up a dozen coke cans and a couple of foxes, which appear as red blotches on the screen and distant pairs of eyes glinting in the distance. 'We've never looked for hedgehogs at this time of year before, perhaps they've gone on holiday,' Mark jokes. It's also quite early in the evening to be looking, he notes, and following a few nights of good weather the hedgehogs might already have full bellies so are less likely to rush out foraging.

But Bryony has suddenly become more focused, fixated on the screen. When we drop our voices, we hear a faint rustling. Mark shines the torch downwards, revealing something small and brown, with a triangular face and oval body, covered in sharp, white-tipped quills designed to protect its softer parts when it curls up into a ball. I can sense Bryony and Mark's relief – both were anxious that we might spend hours wandering around the park and not turn up anything.

The hedgehog continues to bumble along in the most endearing fashion, as if oblivious to the harsh light of the torches and the presence of four very excited humans. 'That's a yearling,' Mark says, 'it's probably never encountered humans before.'

As it is for many species of wild animal, life is particularly tough for young hedgehogs – making it through their first year is a key fight in the battle for survival. If they succeed, then in theory they can live up to ten years, but that's exceptional; most don't make it past their fifth birthday. The research team have put trackers on some of the Regent's Park hedgehogs, embedded in their spines like tiny cubes of cheese on cocktail sticks. They've found that with each year of life it becomes increasingly hard to locate individuals. Since the average lifespan of a population can depend on a number of factors, Nigel is hoping to gather more information over the course of the decade-long study.

Charming as they are, real-life hedgehogs have little in common with Beatrix Potter's highly fastidious Mrs Tiggy-Winkle, who spends all her time ironing and starching clothes. They have some pretty unsavoury habits. This includes rolling on their backs and coating themselves in dog poo, a fetish which scientists have yet to get to the bottom of – perhaps it's a form of camouflage, or possibly an advertisement of some sort to other hedgehogs (maybe to hedgehogs, dog poo smells as perfume does to us, and vice versa. I imagine them twitching their snouts in disgust at my eau de parfum.) The hedgehog diet can also sometimes border on the macabre: two years running Bryony has found hedgehogs chomping away on dead toads. She believes it's not beyond the realms of possibility that they caught the unfortunate amphibians while still alive, which comes as a surprise to me – I've never thought of hedgehogs as nimble or speedy hunters, despite their depiction in Sega's 'Sonic the Hedgehog' video games. But Bryony and Mark claim they are faster than one might think, citing a number of occasions in which a group of volunteers have found themselves unable to catch one. 'The minute you take your eyes off them, they're gone!' Mark insists.

Sadly, this is unlikely to be the case for the hedgehog in front of us – we realise from its shuffling gait that there's a problem with one of its back legs, which my dad points out is dragging slightly. It's a common injury, Mark tells us, which could be down to a fox attack. But he also has suspicions that domesticated canines may bear some responsibility; with urban dog ownership skyrocketing over the past few years it's a rather unpopular observation to make, and a tricky one to prove – a dog might disappear into the bushes and disrupt an entire nest of hoglets,

Mark explains, while the owner remains none the wiser.

This is just one of a cocktail of factors that have resulted in a serious decline in British hedgehogs – it's believed that since the Millennium their numbers have halved in rural areas and dropped by a third in urban environments, highlighting just how poor a shape our countryside is now in. 'I never saw hedgehogs when I was growing up, even though my mum and dad live out in the sticks,' Bryony tells me; 'by the time I was born, all the hedgerows were already gone.' Intensive farming and the use of pesticides has led to a serious decline in the number of invertebrates, hedgehogs' main source of food, and The People's Trust for Endangered Species estimates that between 100,000 and 300,000 hedgehogs are killed on the roads each year. If that's not bad enough, they also get caught up in wire fences and litter, bringing to mind those heartbreaking images of hedgehogs with their heads stuck in the plastic rings that hold beer cans together. Then there's the use of rodenticides – a post-mortem study of over 100 hedgehogs in Bristol found the presence of rodenticides in two-thirds of those animals. When Nigel sent the corpses of twelve Regent's Park hedgehogs off for testing they found detectable levels in eight. Efforts are being made to reduce the use of rodenticides in the park, but there's no easy balance to be found when taking into consideration multiple catering outlets, public toilets and London Zoo.

Finally, there's a general decline in habitat quality and increased fragmentation, which can have a combined effect – animals find themselves stuck on a patch of land that's no longer meeting their needs, with no way out. This is what it's feared could happen to the hedgehogs in Regent's Park, but the team are hoping that by studying them closely they can stave off the problem. 'The park is technically big enough to maintain the population,' Nigel tells me, 'and we want to conserve these animals, but to do it from an evidence base.'

Changes are already under way based on the evidence that's being gathered, as well as broader measures to protect the park's wildlife as a whole, because, as Bryony highlighted in her invertebrate study, it's all part of the same ecosystem. This includes leaving areas of longer grass, thistles and wildflowers to grow uninterrupted, creating connectivity within the park itself. While walking my dog Fox through the park recently I noticed for the first time a crescent of tall grass running from the north-west part of the park near the US Ambassador's residence all the way past the sports fields and right down to the south – a deliberate wildlife corridor. Changes have also been made to hedge management and gaps added to the bottom of fences to make as much of the park accessible to the hedgehogs as possible.

As I watch the little hedgehog we've uncovered scamper off, I'm struck by the fragility of its existence, and yet most people don't even know it exists here at all, on this urban island in the heart of the city. 'It looks mostly healthy,' Mark reassures us, 'I can't see any signs of an infection and it seems to be moving OK.' As it reaches a patch of taller grass, its spines become lost amongst the dark blades, its colours no longer distinguishable in the blackness. Bryony holds up the infrared camera one last time and we watch as the little red blob disappears off the edge of the screen.

SEMI-WILD

In 1637 King Charles I who, according to his chief advisor Edward Hyde, 1st Earl of Clarendon, was 'excessively affected to hunting and the sports of the field', decided to create a deer park. He ordered an 8-mile, 9-foot-high wall of over 5 million bricks to be built around an area of mature grassland, located in what is now one of south London's most affluent outer boroughs; 2,000 red and fallow deer were installed in this new island home.

His actions didn't go down too well with the locals at the time, who were now forced to walk around the entire perimeter to see their families or go about their daily business. Charles ended up paying compensation to those residents who laid claim to the land, and to keep the peace he had to restore the right of his subjects to commute through the park and collect firewood. He wasn't able to enjoy this vast playground for long either; by 1642, the Civil War had broken out. In August 1647 he was able to visit the park again, but under rather different circumstances – by this point he'd been imprisoned at Hampton Court Palace and was now as isolated as the deer.

Following Charles's execution in 1649, Parliament agreed that Richmond Park would be 'preserved as a Park still, without Destruction; and to remain as an Ornament to the City'.

'We're lucky Charles I was the sort of person who fenced off chunks of land he fancied for himself, because otherwise this whole area would be covered in houses by now,' Richmond's Head Keeper, Tony Hatton, tells me, as we drive around the park in his Land Rover on a warm July evening. Tony has been working in the park for twenty-seven years, having started straight out of school as an apprentice. He has an encyclopaedic knowledge of its history and everything in it. It's down to Charles I, he explains, that Richmond is now the largest enclosed acid grassland in Europe. At over 3.5 square miles, it's so big that all of the other Royal Parks could fit inside it, with room to spare. And as a Site of Special Scientific Interest (SSSI) and a European Special Area of Conservation, it makes a huge contribution to the capital's biodiversity.

It's easy to get lost in Richmond Park – if you were parachuted into the heart of this walled island wilderness you'd never know that you were just a few miles outside central London, surrounded by suburbia – and yet there are also plenty of reminders of the park's urban location. An astonishing number of Lycra-clad cyclists speed along the 7.5 miles of road that skirt the park's perimeter. The car parks are forever full. The gates, which close to traffic at sunset, remain open to pedestrians, and from dawn until dusk and through the night, even, the park is teeming with people. From the protected viewpoint at Pembroke Lodge, a Grade II listed Georgian mansion located within the high grounds of the park, you can see right across the city, with St Paul's Cathedral on one side and the turrets of Windsor Castle on the other. It was here that fires would be lit when a beheading took place at the Tower of London to let the Castle know that the execution had successfully been carried out.

Like many urban parks, Richmond is not only an island but a patchwork of different habitats. As well as grassland, it boasts swathes of bracken and large patches of ancient woodland, with over 130,000 trees – predominantly broadleaf species, including oak, beech, chestnut, birch and elm. Rotting timber provides a habitat for a multitude of insects, including the increasingly rare stag beetle, and bats nestle into the gap where the bark breaks away from the tree trunk. There are great spotted, lesser spotted and green woodpeckers, the latter feeding not only in the trees but on the ground, where the land undulates with yellow meadow ant hills, some of which have been there for hundreds of years. Kestrels, sparrowhawks and buzzards hunt by day and tawny and little owls startle those who dally in the park once the light has faded. Deep in the undergrowth, voles, grass snakes and common lizards hide from birds above and mammals below – sharp-eyed foxes, sharp-clawed badgers and sharp-tongued humans. Oblivious to the commotion beneath them, swifts, skylarks, swallows and sand martens zip through the summer air, the sand martens being one of the park's success stories

– their numbers have almost doubled in the space of a couple of years.

Near to the ponds, which were originally dug as reservoirs, Egyptian geese nest in the oak trees, an unusual sight for the uninitiated. Egyptian geese were first introduced in the seventeenth century as an ornamental species. They are early breeders, with their first young born towards the end of February, and a second and sometimes even third brood raised later in the year. I recall in the midst of winter being startled by one of these creatures honking insistently from a tree that had been pollarded. Peering down from this turret, its characteristic brown eyepatches gave it the air of a suspicious pirate.

Just above the surface of the ponds, dragonflies dart and skim, while beneath swim carp, pike, tench and rudd. Kingfishers hunt in a flash of blue, alongside anglers with their curved rods who, unlike the kingfishers, must apply for a permit from the park authorities. Frogs and newts – including great crested – feed on a wide variety of invertebrates. Common terns nest on the square rafts that Tony's team have installed, providing their chicks with cover from predators.

How many of Richmond's 5.4 million annual visitors notice all of this? Not that many, Tony reckons; 'It's amazing how little people are aware of what's around them. They're so stuck on their phones, or busy having a chat, that they're often oblivious.' But no matter how engrossed they are, it's pretty much impossible for them to miss the herds of red and fallow deer, which have roamed freely in the park ever since they were introduced by Charles I nearly 400 years ago. Within a few minutes of our drive across the sun-scorched grass, Tony and I encounter a herd of over thirty red stags, their summer coats glowing reddy-brown in the early evening light. I'm immediately impressed by their size – they're the largest living terrestrial mammal native to the British Isles. I'm also intrigued by their unwieldy-looking antlers, which are covered in velvet and not yet fully grown. Each spring the stags shed their antlers. The older males lose theirs first and lie low for a bit while their younger counterparts still have theirs intact. The antlers regrow over the course of five months; 'It's the fastest growing animal tissue on the planet,' Tony tells me.

While their antlers are growing, the stags are also on a mission to gain as much weight as possible. Then, when the mating season – known as the rut – comes round between September and October, the males use their antlers to battle as the 'master stags' seek to protect their harems of thirty or more female hinds. For the duration of the rut – seven to ten days – the master stags won't sleep or eat and will spend their time roaring and chasing off rivals.

But for the moment, the stags seem pretty relaxed; some are lying down, others grazing, a few stand to attention with heads raised and ears pricked. A couple of jackdaws land on their backs and start picking away at parasites – a service that the deer seem to welcome, according to Tony. 'The jackdaws will go right into their ears and up their noses, and they'll let them,' he notes. In spring the deer even tolerate hair being plucked from their backs for use as nesting material as part of this symbiotic arrangement.

Tony and I are not alone in watching the deer – pretty much everyone stops to look as they pass by. We spot a woman approaching one of the larger stags with something in her hand that she's trying to feed it. Signs across the park clearly advise keeping at least 50 metres' distance from the deer, but few people seem to heed this. On a subsequent trip I spot a middle-aged man attempting to take a selfie with a hind, and a couple placing their toddler within a foot of a stag. I shout at them to move away and they look back, stunned. 'Are they dangerous?' the woman asks incredulously, as if it's impossible a creature living in a London park could harm her child.

The Richmond Park deer are habituated to people, so on the whole they are unlikely to pose a threat – more often than not it's the other way round, humans putting the deer at risk. Tony tells me that he's in the process of trying to figure out the best way to help a stag whose antlers have become hopelessly entangled with an inner tube from a bicycle tyre. He regularly has to deal with deer that have ingested items of rubbish, been hit by cars or attacked by dogs – a Tory MP was recently fined £719 in court after his Jack Russell puppy Pebble caused a stampede during a family walk in the park. 'You should have known better,' the judge scolded.

But during the rut, when the stags are fuelled by their hormones, the risk to humans is greater. A viral video

132

shows one of the larger males headbutting and denting a car, probably in reaction to his own reflection in the windscreen. People have been seriously injured on occasion. 'If you went to Scotland you'd struggle to get within half a mile of the red deer there, because they rarely encounter humans,' Tony explains, 'so Richmond is the perfect place to view them. But they weigh up to 230 kilos, they can run at 35 miles an hour and they have twenty spikes on top of their heads. You don't want to get on the wrong side of them. They could put a few holes in you, that's for sure.' Tony is also concerned that the deer will get the blame for any altercations; 'At the end of the day, it's their home, not ours. They can't go anywhere else,' he argues, 'and we want to keep their behaviour as natural and wild as possible.'

The Royal Parks describe the deer as 'wild animals' on their own website, but in truth, like many urban-dwelling creatures, they exist in a kind of no-man's-land, neither fully wild nor domesticated. Tony describes them as 'managed'. During the winter, they are given a supplementary feed that includes vital minerals which cannot otherwise be found within the confines of the park. This protects them against an irreversible condition called 'swayback', in which the affected deer stagger around as if intoxicated.

'Managed' is also a euphemism for the annual deer culls, which replace the hunts of the past; it's one of the downsides of an isolated urban island population that has no natural predators. Left unchecked, the herd would grow to a size that simply could not be sustained over

133

the winter by the park's available food sources, leaving many animals sick and malnourished, so the herds are kept to around 300 red and 300 fallow deer in total. A gender balance of two males to every five females is also maintained to prevent too many males vying and clashing, with dangerous consequences.

The cull is not something I like to think about, if I'm honest, but Tony – who has devoted his life to the deer – assures me that it's carried out humanely. The park is shut to the general public between 8 p.m. and 7 a.m. while he and his colleagues, who must be very strong marksmen, carefully select up to 200 individuals and put a bullet through each of their heads. The meat is then sold to wholesalers. Watching them grazing peacefully now, I can't help but imagine how the remaining deer must feel when they hear a member of their herd thud to the floor, the smell of blood swiftly reaching their nostrils. Tony claims that neither the reds nor the fallow deer have learnt to fear his team or their guns; mercifully, the memory doesn't seem to stay with them.

So while the deer in Richmond and neighbouring Bushy Park might be wild animals, ecologically speaking they are only semi-wild. But in recent years, wild red, fallow, roe and muntjac – a small Asian species introduced at the start of the twentieth century to Woburn Park which has quickly spread – have been moving further into the city, or island-hopping, if you will. Increasing numbers have been recorded as far into London as Croydon and Forest Hill in the south and Finsbury Park in the north. In a bizarre twist of events, with no apex predators such as wolves or lynx, deer numbers are thought to be at the highest they've been across Britain for 1,000 years. Numbers may have increased even further during the coronavirus lockdowns, during which the closure of restaurants saw demand for venison plummet. It's not known how far into our urban centres the deer will continue to venture (in places like Edinburgh, deer have been spotted in virtually every part of the city), but experts believe that territorial pressures are the likely cause of this urban expansionism.

Tony and I decide to move on, once he's politely but firmly asked the woman feeding the deer to stop doing so. We pause briefly at one of the ponds, where the stillness of the evening is interrupted by a persistent buzzing – not a grasshopper or cricket, a fly or a bee but something altogether more irritating, a drone. (Even more irritating, the residents of Richmond upon Thames would argue, are the aircraft that thunder above our heads every few minutes on their way to Heathrow.) In these ponds, Tony tells me, he once recovered a giant snapper turtle – one of the many abandoned pets he's encountered in the park over the years. The prehistoric-looking creature, which he bent down and, perhaps unwisely, picked up with his bare hands, weighed over 25 kilos, with a head the size of a tennis ball. It's now living out the rest of its days at the London Aquarium.

Back towards the gate where we started, we spot a group of fallow deer next to the road, grazing an area of long grass, dotted with patches of yellow ragwort. They seem even less fazed by our presence than the reds (they have even less road-sense too, Tony adds). Like the group of reds we previously encountered, all are male – known as bucks in fallow deer; the females are does – but they are considerably smaller and more delicate. Long eyelashes fringe their hazel eyes and their white-spotted chestnut coats give them a soft and gentle appearance. Whereas the reds' antlers resemble the branches of a tree, the fallows' are palmated, like a cartoon hand with fingers extended.

We watch as one of the bucks extends up on its hind legs, plucking a bunch of leaves from a tree with impressive dexterity. You can tell where they've been, Tony says, by what's known as the browse line, which stops at the point where they can no longer reach. 'They're fussy herbivores,' he says. 'They'll munch for a bit then move on.' In the autumn they'll shift their attention to the acorns and conkers as they fall from the trees, and in winter they'll harvest the brambles. Like many ruminants (herbivores with four-part stomachs that are used to ferment a plant-based diet, prior to digesting it), their life expectancy is closely correlated to the speed at which their teeth wear out. They can survive longer in lowland environments like this one, where soft grass makes up a large part of their diet.

As the light starts to fade, Tony stops just before we reach the gate, parking up next to a fenced-off enclosure that's sometimes used for cattle. It's currently left open for the deer to use as a place of refuge away from the crowds when they need it. A little owl lands on one of the fence posts and turns its head towards us, peering out through huge, saucer-like eyes. Tony kindly drops me at Richmond Station on his way home and I'm back in central London in less than thirty minutes.

Three months later, in the first week of October – peak rutting season – I head back to the borough of Richmond, this time to Bushy Park. At 1,100 acres, it's the second-largest of the Royal Parks after Richmond. I take my dad along for company and, neither of us being familiar with Bushy, lead us round in circles for a good half-hour before asking a ranger for help locating the deer. After reassurance that we have no plans to get within fewer than 50 metres of them, he advises us to head to Heron Pond at the centre of the park.

Before we even reach the deer, our attention is drawn to a peculiar spectacle: the park has been descended upon by dozens of humans, mostly male, clutching cameras with huge telephoto lenses. Some are wearing head-to-toe camo gear, which is doing little to camouflage them as they stand next to the car park sipping disposable cups of coffee.

A group of these men surround an area of bracken where a single stag is standing. He is different in appearance to the males I saw in Richmond Park in the summer – he's clearly been piling on weight for the last few months, and his coat has grown rough and shaggy, particularly round his broad, muscular neck. He has a scar on his face near one of his eyes and looks like he's been through the wars. Lifting his head, he reminds me of a walrus as he opens his mouth improbably wide, letting out a deep, roaring, burbling, belching sound, accompanied by dozens of camera shutters – click, click, click.

I suddenly notice a group of females lying in the scrub behind him, incredibly well hidden for animals of their size. Two men and a woman stroll along a path cut through the bracken just a metre or two away, oblivious either to the deer or to any potential danger. The photographers, some of whom are also standing perilously close, locked in a show of apparent machismo with the stag, reframe their shots to maintain the illusion. (They have already been careful to crop out the car park and café selling teas, coffees and a range of hot and cold sandwiches.) One photographer pauses to proudly show me an image he captured just the day before, with a pair of young males locking antlers – he wasn't sure whether the fight was real or just play, he tells me.

From across the car park another stag hollers, but neither makes any bold moves. It seems to me that, menacing as they might seem, most of the stags' time is spent merely posturing, like crusty old politicians in the House of Commons. We keep our distance as my dad, who is relatively nonplussed by all this bravado, watches a green parakeet poke its head out of a hole in a nearby tree and points to a black-headed gull in winter plumage, standing on a little island in the centre of the pond. After a short while, the stag nearest to us starts to slowly move and the men in camo gear scatter in fear. We're left considering the strange irony of these huge beasts; while most people don't consider them wild by virtue of their urban island location, here are dozens of amateur photographers scrambling to depict them as such in a photo they can proudly share on social media.

SQUIRREL ISLAND

On a warm day in early summer, as I'm sitting on a bench in our garden, watching the gang of resident sparrows squabble at the bird feeder, my phone pings. It's a message in a group chat I have with two of my close friends from school, Ceri and Leah. 'We have to pay a man to murder some squirrels for the simple crime of living in our attic and eating our electrical cables,' Ceri writes, 'and this feels like a part of adult life – paying someone to kill cute animals – that no one warned me about?' Leah and I both express our sympathies. 'Can't he just remove the squirrels and release them somewhere less cable-y?' Leah asks.

The answer, sadly, is no. From December 2019 it was made illegal for anyone to release grey squirrels, once captured, into the wild in the UK. This includes squirrels that have been rescued and rehabilitated: by law, rescue centres now have to euthanise any grey squirrels that are brought to their doors. The UK government has justified the decision by claiming that grey squirrels threaten native wildlife and harm the economy, causing damage to timber through their habit of stripping bark from trees (and presumably also due to their penchant for munching electrical cabling). This has, unsurprisingly, caused a bit of a stir and met with condemnation in some quarters. In an article published in *The Conversation* at the time, Jason Gilchrist, an ecologist at Edinburgh Napier University, argued that there are 'glaring contradictions in what is considered invasive and what needs to be controlled', highlighting one such contradiction, the pet cat, of which there are 11 million in the UK. Cats kill an estimated 27 million wild birds and 92 million wild prey in total each year. 'Why is it acceptable for animal shelters to rescue an invasive alien species, the domestic cat, and for the public to allow them to roam free,' he demanded, 'but unacceptable for wildlife rescue centres to help and release a few grey squirrels?'

Grey squirrels were first released in the UK in 1876 by Victorian banker Thomas V. Brocklehurst, according to the first verifiable records. Brocklehurst brought a pair of the exotic rodents back from his business trip to the US and let them loose in his garden in Cheshire. (It seems this predilection for exotic species ran in the family – in the mid-1930s, his relative Sir Henry C. Brocklehurst went on to release red-necked wallabies into the Staffordshire countryside.) Brocklehurst's friends and neighbours liked the look of his squirrels so much that they began acquiring their own, and a wave of grey introductions followed in the early twentieth century; in London 100 were released in Richmond Park in 1902 and 91 in Regent's Park between 1905 and 1907.

These days, grey squirrels are one of the few urban

species that almost everyone can name. They're not, somehow, as invisible as pigeons, which regularly go unnoticed on our city streets, as if simply part of the brickwork. Nor are they as universally disliked as their rodent ancestors, rats and mice. Children, in particular, are enchanted by them.

Their success in urban environments is down to a few factors. The urban heat island effect, combined with access to food sources all year round through bird feeders and discarded rubbish, means that they are more likely than their rural counterparts to birth a second litter each year. As omnivores and generalists, there isn't much they won't eat. They have the ability to learn and to grow accustomed to people, gatecrashing picnics and snatching food directly from the hands of humans that feed them. They do still require some access to more conventional green space – the squirrels that dash across the roads and pavements in the close where we live will quickly escape to a tree in someone's garden the moment they are startled. But, as we already know, our cities are full of green islands.

There is, however, one big issue that many people have with these fluffy-tailed, bright-eyed, tiny-handed immigrants, and that is the impact they've had on our native red squirrels. It's an impact so great that in my lifetime I've never seen a red squirrel in the wild. For me, as for much of the British population, red squirrels – with their rich auburn fur, cute button noses and adorably large tufty ears – are confined to a world of television programmes, magazines and Christmas cards. They are, in fact, almost as exotic as the grey squirrels of America were to Thomas V. Brocklehurst in the nineteenth century.

'If we hadn't introduced grey squirrels, there would be red squirrels everywhere. Probably even in London,' squirrel expert Kat Fingland tells me. Kat spent her PhD studying one of the UK's few populations of urban red squirrels in Formby, Merseyside (the other notable example being in Aberdeen). True, grey squirrels tend to be a bit bolder. 'If you put up a feeder,' Kat says, 'grey squirrels will run up and start using it straight away, whereas reds may take a few days and approach with more caution. But once they get their head around it and realise it's not going to attack them they'll happily use it too.' Reds are found in cities across Europe, so we have no reason to believe there's anything other than the greys standing in their way.

The relationship between grey and red squirrels is often misunderstood – it's a common misconception that greys kill reds. They don't. There's no deliberate ploy on the part of the greys to wipe out their red counterparts, like two rival gangs in a turf war. In fact, researchers have noted that if reds and greys find themselves on the same feeder, the greys will chase off other greys, the reds will chase off reds, but the two species will largely ignore one another. The trouble is that, coming from America, where a number of different squirrel species cohabit, greys have evolved competitive advantages that the reds have never encountered before.

The two species eat a similar natural diet (as opposed to the kebab my friend Laura once witnessed a grey squirrel and a pigeon fighting over in Tottenham). In areas of coniferous woodland red squirrels fare a little better, because the seeds of pine cones alone are too small to sustain the larger greys, but a mixed woodland is best for both, with a variety of species such as sweet chestnut, beech, birch, oak, pine and rowan providing a reliable source of food all year round and enough diversity to protect against different tree diseases which can wipe out single species. Grey squirrels weigh a third more than reds, so they eat a lot more, and unlike the reds they are able to digest the phytotoxins found in immature acorns. This means that they will eat an entire acorn crop before the reds – who have to wait until the acorns ripen – can even get a look in.

On top of such fierce competition for their dinner, the reds face another threat – squirrel pox. Grey squirrels carry squirrel pox asymptomatically, but if the reds catch it most will die within a matter of weeks. The issue is compounded in urban areas. 'Adaptable species that can succeed in urban environments tend to live in higher densities,' Kat explains, 'so, for example, there are higher densities of urban foxes than rural. And it's the same with red squirrels – the density in Formby is ridiculously high. But this success can be their downfall sometimes too.' The virus can quickly spread through these dense populations, which are then liable to crash – in 2008, an outbreak wiped out about 80 per cent of Formby's red squirrels, leaving just tens of individuals. In 2019, after the population had rebounded, another wave reduced it again by a half. This can also impact genetic diversity, with populations ultimately becoming unviable.

I can't help but draw comparisons with the deadly coronavirus pandemic, which is still very much under way across the globe at the time of my conversation with Kat. It all feels a bit close to home. The similarities aren't lost on her either. While scientists were battling against the clock to develop an effective COVID-19 vaccine for humans, researchers at the Moredun Institute in Midlothian, Scotland were working on a squirrel pox vaccine. There is also a hope that the reds might gain some kind of natural herd immunity. In 2008, 8 per cent of the population were found to have gained resistance to squirrel pox. Over time, Kat hopes that this number will grow, but it's a slow process.

A few weeks after speaking to Kat I'm on my way to Formby in the hope of meeting some urban red squirrels for myself. Ben and I leave the house at 6 a.m., to his significant displeasure. ('You said we were going on holiday!')

Formby is both a commuter town for Liverpool and – in the summer months – a tourist destination, popular for its beaches and sand dunes. We park at the spot where I've arranged to meet with Rachel Cripps, the regional Red Squirrel Officer (her official job title), by a fringe of imposing coniferous woodland that separates the town from the sea. From here you can hear the gentle lapping of the waves, while the trees cocoon you from the wind and the wilder elements. The woods, and the sand dunes to which they give way, are filled with delicate wildflowers in complementary shades of yellow and purple. There's lemon-hued evening-primrose, and ragwort, with its flowers that look like tiny suns, petals bursting out from round centres like rays. There's common stork's-bill, with delicate lilac flowers, purple harebells, like tiny paper lanterns, and small bugloss, with hairy leaves and tiny flowers resembling those of its botanical relative, the forget-me-not.

Rachel visited Formby in her final year of studying Zoology at Liverpool University, and she dreamt of one day working here. After spending a few years at the RSPCA and completing her Masters she landed the Red Squirrel Officer job, and now covers an area spanning 197 square kilometres with 142 areas of woodland, from Crosby up to Southport and into West Lancashire. 'I can't believe it – I dreamt of it, and it came true!' she declares with absolute earnestness.

When she first started, the role was focused on grey squirrel control – a slightly controversial subject – but since then it's expanded to include coordinating all the conservation activities associated with red squirrels; working with a team of volunteers on a monitoring programme and surveys, engaging with the community and delivering public awareness campaigns. Formby is a conservation area, meaning that a licence is required to fell any trees, but this still needs close monitoring, according to Rachel. She also encourages wildlife-friendly gardening, such as planting fruit trees and maintaining hedgerows, and advises residents who like to feed the squirrels on what kind of food to provide. This includes a mixture of fruits, linseeds and sweet chestnuts as well as peanuts and sunflower seeds, not forgetting fresh water, especially in summer.

Rachel has brought her squirrel-detection dog along – a sweet-tempered German Shepherd called Max. Max was initially trained to detect drugs, explosives and pyro-technics, she explains, but after a firework was thrown at him he became afraid of loud noises, disqualifying him from the job. So Rachel trained him to sniff out red squirrels instead. Apparently he's very effective at this, uncovering sickly or already deceased reds hidden deep in the undergrowth in minutes, helping Rachel and the team to control the spread of squirrel pox.

Formby shows what could, or should, have been when it comes to urban red squirrels. Rachel has chosen the woods as our first port of call because we're pretty much guaranteed a sighting here, but they can be found dotted throughout the town itself. Google Formby and images of red squirrels will immediately pop up – they are clearly deeply loved in the town; houses, pubs and even roads are named after them. 'We regularly hear residents saying, "I've spent £50 on peanuts for the squirrels this month,"' Rachel tells me. The squirrels build their dreys – spherical nests with a hole in the middle made of twigs, moss, feathers and whatever else they can get their hands on – on top of roofs or in people's gardens if the trees are tall enough. Within the town they face fewer predators, such as goshawks and other raptors. And, unlike purely terrestrial mammals such as hedgehogs, the squirrels can use the town's tree-lined streets as green corridors to get around, leaping from branch to branch with acrobatic precision. According to Kat's research, they've become pretty savvy too – where possible they seem to avoid crossing the road at street level, preferring to use the tree canopies to get from one side to another.

Rachel, Ben and I begin our search on the edge of an area known as 'Squirrel Walk'. Until recently the National Trust, which owns this area of woodland, fed the squirrels via metal feeding platforms installed high up in the trees. But this practice was stopped following the latest outbreak of squirrel pox. (Formby's residents were also asked to take down their garden feeders to avoid the spread of the disease.) So now, the squirrels have spread out beyond the pine trees of Squirrel Walk, seeking a greater diversity of natural food sources. Rachel hopes that they'll be able to start using the feeders again soon by hanging fruits – apples, pears and plums – from the branches, which would pose less of a risk.

According to Kat's research, the squirrels that live in the town centre tend to stay there, while the squirrels that live in these woods rarely venture into town, but both have become accustomed to the presence of humans. The National Trust used to sell peanuts at the gate, which the reds would willingly take from people's hands. This prac- tice stopped some time ago, once it was recognised that hand-feeding wildlife isn't the best idea, but the squirrels will sometimes still approach expectantly, Rachel tells us.

A few minutes into our walk we pause as we hear the sound of organic matter lightly raining on to the spongy forest floor. Glancing down, we see evidence that points towards the source of the sound; amidst the pine needles are the cores of pine cones, the scales torn off by what must be extremely strong, sharp little teeth in order to expose the nutritious seeds. As we strain our ears, we're able to detect a faint munching. This is swiftly followed by a much louder 'tschuk, tschuk, tschuk' sound above our heads as, having now spotted us, the red squirrels begin stomping their feet in response as they climb up the tree trunk. 'They're saying, "I know you're here. I'm watching you!,"' Rachel laughs. Silhouetted against the tree canopy, I can immediately see that they are considerably smaller and more delicate than the grey squirrels I'm accustomed to. The weather is pretty overcast, but a few breaks in the dense foliage allow us to make out their colour as we get closer; one is almost completely ginger, the other dark brunette, with a tail that's almost black. Both have characteristically tufty ears.

Stomping aside, the pair seem relatively comfortable in our presence, and in the presence of Max the German Shepherd too. Their behaviour is very different from that of the reds just a few miles up the road, Rachel explains, which always keep their distance. It's hard not to anthropomorphise, watching them use their tiny hands to grip the pine cones. When I spoke to Kat, she admitted she found herself doing the same thing; through her experience of catching and tagging reds over the years she believes that some are shy, others bolder. Some seem to storm off in a huff when she releases them, affronted at having been caught in the first place. 'Kat has been looking at the flight response of the squirrels here,' Rachel tells me, 'and how they behave when they come into contact with people in different environments. We reckon our squirrels are the friendliest red squirrels internationally!'

After a few minutes the squirrels move on and so do we, as we continue our walk to the edge of the woods. The ground becomes sandier and I start spotting increasing numbers of limpet and mussel shells. Soon the trees give way to dunes, dotted with patches of grass and brush, and then the sea itself. During dry spells the squirrels will sometimes venture out, Rachel says, leaving the cover of the woods in search of pools of rainwater in the dunes. But beyond this point they are effectively cut off.

It's this natural barrier – the sea itself – that allows Rachel and her team to support the reds in Formby. With the Irish Sea to the west, the team have formed a buffer zone between red and grey squirrel territory along the eastern side. They have, in effect, created a red squirrel island. This is a technique, Kat tells me, that's being used in areas across the UK. 'Island populations – such as the population of reds on the Isle of Wight – are easier to protect,' she explains, 'but in Formby, at least you only have to defend the stronghold on one side. The aim is to keep pushing the greys back, thereby expanding the area where the reds are succeeding. There are similar projects ongoing in other places, including plans to introduce reds to parts of Cornwall, for instance.'

'Pushing the greys back', it should be noted, is another euphemism for culling. The reality in Formby at least is that grey squirrels are trapped wherever they're found in live capture traps (in case a red is trapped by accident) and then humanely killed, either with a shot to the head or 'cranial dispatch' – a sharp blow to the back of the skull. If you're feeling squeamish about this, you're not alone. For those of us who've grown up surrounded by grey squirrels it's particularly hard to stomach. People develop an affinity with the wildlife that's around them and that connection with nature is important – it's what makes us want to protect it. For most urban dwellers, grey squirrels are one of the few wild mammal species we regularly encounter, so we're naturally going to baulk at the idea of destroying them.

On the other hand, research shows that people who have grown up with red squirrels and have seen them suffer over the years are much more likely to be in favour of culling the greys. Having the support of local residents is an essential part of the conservation work that takes place in Formby. 'You need a certain percentage of the human population to be on board to even stand a chance,' Rachel admits. She relies heavily on locals reporting sightings of grey squirrels so that they can be trapped and killed before they reach the woodland, where they would breed in large numbers and things could quickly spiral.

For most people, it's not entirely black and white – some will be so eager to protect the reds that they'll trap and kill the greys themselves, Rachel tells me. Others won't want to deal with 'the dispatch side of things', but they'll agree to having a live trap in their garden, or to report any grey sightings to the team. The whole thing gives me pause for thought. Despite understanding the importance of Rachel's work, I'm not sure I myself could pick up the phone to sign what is effectively a grey squirrel death warrant.

A more palatable solution, perhaps, would be to use an immunocontraceptive that only affects greys, which could be added to feed left out for the squirrels. Such a contraceptive is already in development, and – since trapping is extremely time-consuming and may ultimately be a losing battle – could also prove a more sustainable and cost-effective way of managing the greys.

Still, I feel I must put the question to Rachel: in an urban environment, where it's difficult for any wild species to survive, should we really be destroying those that are successful? 'If we weren't controlling grey squirrels there would be no reds left,' she replies; 'they'd already be gone. If you didn't do any kind of management, particularly in urban environments, you'd end up with only the most gen-eralist and competitive species – grey squirrels, pigeons, magpies, rats and mice . . .' As is so often the case when it comes to urban conservation – and, indeed, conservation in general – the solutions are not simple, nor are they always pleasant. Being common and successful is enough to put a target on your back, as the rights and claims of one species are weighed up over another. I admire Rachel and her team immensely and she's rightly proud of the work she does on behalf of Formby's red squirrels. I'm just glad I'm not the one who has to make these decisions.

In the last few years, a team of international scientists in the Galápagos have once again been measuring the size of finches' beaks; specifically, the beaks of medium ground finches. In many ways, Galápagos is the archetypical pristine island wilderness. My father and I often comment that, had I not taken so many photos, we might believe our visit was purely a dream. But, otherworldly though it is, Galápagos is not as free from human presence as many might think. In June 2019, there was an outpouring of horror and indignation when it was announced that US aircraft would start using the airport on San Cristóbal island as a military base. And with a population of over 25,000, plus an influx of 150,000 tourists each year, Galápagos was already far from being a place of sanctuary for its wild inhabitants alone.

Most of Galápagos's human population lives in the town of Puerto Ayora, on Santa Cruz, which boasts all the amenities you might expect from one of the world's biggest tourist destinations; and it's not just the tourists who have taken to frequenting Puerto Ayora's burgeoning restaurant scene. The local finches have also become regulars. In fact, I remember one landing on a red-and-white checked tablecloth right in front of my nose while I was polishing off a pizza with my dad during our visit. They are unassuming little birds, not technically finches at all, but either tanagers or buntings – ornithologists aren't quite sure. With their small stature and relatively drab plumage they could quite easily fade into the background, if it weren't for their unusual behaviour. Like most of Galápagos's charismatic wildlife, having only encountered humans for a very short length of evolutionary time, they exhibit a disarming fearlessness.

Researchers have found differences between the beaks of these urban finches and those of their non-urban relatives, and they believe that diet may have something to do with it. In the past, the finches' beaks were shaped by selection pressures, which favour either larger or smaller beaks, depending on the availability of different types of plant seeds. But the medium ground finches in Puerto Ayora seem to have developed medium-sized beaks, suited to the easy pickings of crisps, biscuits, bread and ice cream that humans leave behind.

What's more, while finches outside the city seem to show little interest in humans and their food, the researchers have found that the Puerto Ayora finches will respond excitedly to the rustle of a crisp packet.

Even on the world's most remote islands, urbanisation is impacting the way that wildlife behaves and, incredibly, its evolution. Meanwhile, a study much closer to home has found that we Brits may be similarly shaping the beaks of our native great tits, through our love of feeding them when they visit the green island gardens attached to our homes. According to a research paper published in the journal *Science*, the British spend twice as much on feeding birds as our counterparts in mainland Europe. The study looked at the DNA of over 3,000 great tits in the UK and the Netherlands and found that the British birds have developed beaks that are 0.3 millimetres longer than those of great tits across the Channel. The researchers believe this helps them make the most of our garden feeders. If the appetites of my resident great tits are anything to go by, it seems entirely plausible.

These changes have happened over a short space of time; according to scientists at Oxford University, who have been recording beak measurements in nearby Wytham Woods for seventy years, there's been a substantial increase since the 1970s. Those with longer-beak genes appear to be more successful at reproducing in the UK, which could also explain why we've been seeing an increase in great tit populations over the past few decades.

As a product of living cheek-by-jowl with humans, who transform the landscapes around us at unprecedented speeds, the creatures that inhabit our urban islands are forced to evolve and adapt at a pace that Charles Darwin himself could never have predicted. The decisions which those who seek to protect our urban wildlife are forced to make are often complex and messy. They must weigh up the interests of one species over another (and all too often it's the interests of our own that win out). There's so much more to be done to make our urban islands more welcoming and more accessible to nature. Thoughtful planting, protected areas away from humans and hounds, green corridors to connect these archipelagos, and sometimes simply letting things grow, are just a few places to start. But more on that in the final chapter.

For now, it's time to find webbed feet, to grow fins and sprout tails. Because our cities are full of water and we're about to dive in.

CHAPTER 4

Look In: Hidden Rivers and Watery Realms

Ever the river has risen and brought us the flood,
the mayfly floating on the water. On the face of
the sun its countenance gazes, then all of a sudden
nothing is there.

Anon, *The Epic of Gilgamesh*

On a Sunday afternoon in early May a heron stands on the bank of a pond, perfectly balanced on one leg, its grey wings folded and head hunched like an old man in a tattered cloak. On the other side of the water, another of its species appears to be without any head at all, so well tucked is it into the bird's downy breast. A pair of Egyptian geese stretch their long necks, plucking at the shoots of a huge weeping willow, while a moorhen pecks absent-mindedly in the gravel nearby.

A few metres away, a crowd of humans surrounds the water, parents chatting while their children run about chasing one another. Things remain relatively calm for a while, but the birds are on private property and the geese will soon be reminded of such.

On the sidelines, a squadron of eight Humboldt penguins assembles, waddling into position. Slightly smaller in size, but greater in number, they surround and goose-step the unfortunate waterfowl to the far corner of the pond, where they are left with no choice but to retreat into the water.

159

The penguins pause by the pond's edge, white chests puffed out in a show of apparent triumph, before returning to their regular penguin-based activities.

This mixture of native and exotic species is just par for the course of urban life, where creatures, both human and other animals, become so used to encountering new things every day that novelty itself loses much of its strangeness. Though increasingly common in London, the Egyptian geese themselves are immigrants, having escaped from ornamental lakes and ponds before establishing themselves across the south-east of England.

London Zoo's Penguin Beach is an odd refuge for aquatic wildlife, but not the only one in the capital. Nearly a year earlier, I stood and watched a female mallard, trailed by nine tiny ducklings, splashing about in one of the Trafalgar Square fountains. It was a blisteringly hot day in the midst of the first coronavirus lockdown, so the square was empty of tourists, but a few passers-by out doing their daily exercise stopped to take photos on their phones. The resident security guard told me that he often saw birds in the fountain, but never before a whole family.

How the mallards first arrived was a bit of a mystery, seeing as the sides of the fountain were too steep for the flightless ducklings to get in or out. The guard was convinced someone must have put them there, but it's more likely that the mother constructed her nest inside the fountain and when the ducklings hatched they simply plopped in. Either way, the water being chlorinated and the fountain affording no food and precious little shelter, it was clearly no suitable habitat for the birds, so I made a call to a local wildlife service as I left and a group of volunteers removed the ducklings to safety.

These episodes show the opportunistic nature of many of our waterbirds, with cities becoming home to a rising number of species such as cormorants, coots, herons and even great crested grebes. But the truth is, wherever you find water you'll find wildlife in our cities; from the bath-tub my neighbour left in her garden, which quickly filled with algae and mosquito larvae, to the River Thames, where humpback whales, dolphins and even a beluga whale have turned up at one time or another.

Green spaces have become a familiar concept to city dwellers, but our urban blue spaces – bodies of water, big or small – are full of life, as we will discover if only we take the time to look.

A TALE OF TRANSFORMATION

It was water that originally drew our own ancestors to the site of most of our modern cities. In London from around 13000 BC onwards, early settlers built their camps close to the rivers, springs or lake margins. By 4000 BC, at least one settlement was growing crops along the banks of the Thames, evidenced by a cache of burnt wheat grains found near Canning Town. But it wasn't until around 2000 BC that humans began to make a dramatic mark on the landscape around the Thames and its tributaries through agriculture. As farming gradually superseded herding, the land was systematically cleared and divided into hedged fields.

'It's a combination of the smaller and the bigger rivers that brought people to this place,' writer, walker and author of *London's Lost Rivers*, Tom Bolton, explains when we meet for a stroll along the River Wandle in south-west London. 'You've got the big tidal Thames that's great for access and transport, but then the smaller rivers like the Wandle, which are more useful for power and drinking-water, and everything else besides.'

The impact modern humans have had on our waterways and waterbodies cannot be overstated. We've polluted, dammed, redirected, straightened, drained and culverted. Getting rid of a river altogether isn't easy, so in many places we've simply built on top – throughout our cities the ghosts of forgotten waterways still flow, hidden beneath our feet. 'You can look down the covers of storm drains throughout central London and see a river flowing,' Tom describes; 'you can even hear them sometimes. These watercourses are the only real evidence we have of what was here before us – they go right back to the natural landscape.'

Tom has spent many hours researching these lost rivers, which no longer appear on maps but can be traced through their impression on the land from high to low ground. The Wandle itself is buried under central Croydon before briefly emerging in a park further north-west, and then again from under an industrial estate in Beddington. It

162

continues up to Mitcham, through the National Trust-owned Morden Hall Park and up to Colliers Wood, the stretch where Tom and I are taking our walk (the name Colliers Wood perfectly encapsulating both the natural landscape that used to exist in this area and the practice of charcoal-burning that displaced much of it).

Up until the late 1990s, the Wandle was a casualty of the booming industries that developed along its banks. Initially used by the Normans as a power source for their mills, in medieval times the Wandle hosted trades like malting and tanning, followed by factories and warehouses creating gunpowder and military equipment. From the eighteenth century onwards it became the home of dye, paper and silk manufacturers. Many of the products created here were designed for specific purposes, including rainproof pigments to colour the hats of cardinals in Rome, felt for piano hammers and, later on, fine leather for the seats of Ferraris, Lamborghinis and Aston Martins.

All of these industries discharged large volumes of waste directly into the river and it became a familiar saying that you could tell what day of the week it was from the colour the Wandle was flowing.

By the 1960s, the Wandle was officially declared a sewer. 'We see pollution as a modern problem,' Tom observes, 'but this is an industrial river going back to the eleventh century and probably earlier – there's never been a clear separation between untouched nature and human activity here. It's always been a combination of the two, trying to coexist.'

And yet today, as Tom and I stand on a bridge in Wandle Park looking down at the river flowing beneath our feet, the water is surprisingly clear – we can see all the way to the bottom. Grasses and lush green vegetation erupt from the banks and a blackbird serenades us with gusto from the undergrowth.

Now described as 'one of the UK's ten most improved rivers' by the Environment Agency, the Wandle's story is one of transformation. It remains a decidedly urban river – just outside the park, we watch as it rushes past a huge Sainsbury's, sandwiched between two busy A roads. But even here the improved health of the river is evident. The sunlight glints on the surface of the water and thick fronds of green weed are swept to and fro by the current, like tresses of hair in the breeze. According to a group known as The Wandle Fishermen, who regularly fish here, this former sewer now contains mirror and common carp, barbel, koi, brown trout, chub, gudgeon, roach, dace and the critically endangered European eel. Herons and kingfishers breed here and little egrets make regular appearances too, Tom says.

It was dedicated groups of environmentally minded volunteers that helped turn the tide for the Wandle. Then, in 2001, The Wandle Trust was formed, an environmental charity that still holds community river clean-ups on the second Sunday of every month, hauling out everything from shopping trolleys to shotguns. But this success was only made achievable once the industries that had so badly polluted the river began to decline, coinciding with the introduction of new EU directives to improve water quality. And it was these directives that triggered a major clean-up of many waterways across Britain.

Another major beneficiary was the Birmingham canal system, which was designed for transporting coal, iron and other bulky materials, and which Ben unwisely swam in while at university in the city, contracting a nasty bout of shingles soon afterwards.

'These canals would've been the most polluted, horrible places you can imagine during the heart of the Industrial Revolution,' Paul Wilkinson, Senior Ecologist at the Canal and River Trust, tells me in his Black Country accent.

I meet Paul in late April on a stretch of towpath in Birmingham city centre, outside the rather incongruous Sea Life Centre. Three mallards and a Canada goose snuggle together on a brick island: a turn junction – or aquatic roundabout – where two canals meet, with a signpost pointing towards Wolverhampton in one direction and Fazeley in the other. Were it not for the water, this area would be entirely encased in hard, impermeable surfaces – bricks, paving, concrete.

Like the Wandle, Birmingham's waterways used to be devoid of any life, save for mosquito larvae, which seem to survive anywhere. 'They were basically seen as drainage ditches,' Paul states; 'all the local industry would use them to get rid of their dirty water.' But things changed in the 1990s when the directives came into force, and considerable investment was put into cleaning up these canals. This involved a lot of dredging – scooping out and disposing of the heavily contaminated sediment at the bottom – as well as the removal of rubbish and other remediation works. Plankton soon began to appear: tiny crustaceans such as daphnia, like miniature rowing boats, and microalgae known as diatoms, now known to be responsible for 20 to 50 per cent of the planet's oxygen supply.

Improving the water quality was just the first step. Since then, along this stretch of canal alone Paul has been working with over fifty volunteers on a rewilding project, using a vast assortment of plants as diverse as the city itself to fill each and every available patch, niche, gap, crack or crevice they can find. 'We've got lock cottages along the canal network that had 200-year-old gardens where things would've spread out and escaped along the canals,' he explains, 'so we've tried to amalgamate those sort of plants with native species, as both are good for wildlife. I want to trigger people's memories of their nan's garden with smells of mint, sage and lemon balm. I want to slap them in the face with some nature!'

As we meander along the towpath, Paul is constantly on the lookout for new spots to plant. He proudly points out the patches of green where the team have been working – purple lavenders, yarrow and pink-petalled mallow, sea kale, with its frothy white flowers that smell of vanilla, custard-coloured primroses, bluebells and wild garlic. The plants have taken on a life of their own; oxeye daisies, lemon balm and lamb's ear spill out of the beds and sprout from cracks in the walls next to climbers – honeysuckle, jasmine, wisteria and grape. The urban heat island effect results in temperatures that are up to 5 degrees higher here than outside the city, allowing the team to grow things that usually thrive in warmer climes. Paul's latest project is to add some banana plants; 'It's a blank canvas – you wouldn't be able to do that in a SSSI, now would you?' he grins.

For the most part the topsoil, which was covered in silt when the canal was dredged, is too nutrient-rich for wildflowers. But in one patch Paul is attempting to create a tiny wildflower meadow – if the local Canada and greylag geese will allow it. 'They eat everything we put in,' he sighs, 'and because they don't migrate they're always here!' In an attempt to dissuade them the volunteers have erected a tiny fence around the meadow, which, surprisingly, seems to be showing signs of success. Wild carrot, yarrow, viper's-bugloss and cowslips are already springing up, along with yellow-rattle, wormwood and knapweed, its purple, thistle-like flowers attracting clouds of pollinators in the summer months.

Slightly further along is Paul's most ambitious project – a linear orchard that will run along the towpath all the way from Wolverhampton to Worcester. Although the trees are only a few years old, some have already grown taller than the fence behind, their branches covered in soft pink and white blossoms. In autumn people can help themselves to the apples, pears, plums, cherries, crab apples and peaches, Paul tells me; 'The fruit rarely lasts for long, especially the peaches.' The fruit trees will also play an important part in supporting local wildlife, with a whole host of invertebrates feeding on the flowers and leaves, while birds snack on the fruit.

Paul has deliberately chosen to introduce only hardy plants that shouldn't need much in the way of upkeep, but he stops to show me a few species that have been here of their own accord since before the canal was built over 200 years ago, when this whole area was farmland. Greater celandine, with its waxy yellow flowers, groundsel and shepherd's purse spring up along the towpath. Even within the hard, brick-lined sides of the canal itself sprout patches of skullcap – a self-seeding plant with violet tubular flowers that feeds the yellow skullcap leaf beetle.

With the exception of skullcap, few plants are able to survive in the canal unaided, since it lacks the gently sloping muddy sides of a natural waterbody. But every 100 metres or so, at around 2 metres deep, the Canal and River Trust have lashed bundles of hazel wood to the brickwork, topped with coir rolls made from coconut-shell fibre. This bioengineering solution allows the volunteers to add water plants; when the purple-loosestrife and yellow flag irises come into bloom they'll be covered in bees, dragonflies and damselflies, Paul tells me. He's used a similar technique to create a little basin, with the addition of reeds, white water-lilies and marsh cinquefoil, with its unique, star-shaped magenta flowers. 'I'd love to be here when a botanist walks past one day,' he jokes; 'they'll think, "What on earth is that cinquefoil doing there!"'

In a busy section of the canal surrounded by bars and

restaurants, Paul has been experimenting to see whether the water is clean enough for aquatic plants to survive. People still dump stuff down drains that connect directly to the waterways, and rubbish is a continuous problem. But thus far the plants appear unimpeded; dense clumps of yellow flag irises and water mint now shield a row of wheelie bins, testament to Paul's success. Once established, the plants can contribute to the clean-up efforts, their roots sequestering oils and pollutants from the water.

The canals flow into the rivers, and the rivers into the sea, so the health of these waterways has far-reaching consequences. Where the conditions allow it, they provide wildlife with connected habitats and aquatic species are quickly able to move in. 'Aside from perhaps the railways, there's no other corridor that connects the countryside to the city in the same way,' Paul argues. 'You can walk unhindered all the way to Worcester, to Cannock Chase, to Wolverhampton without barriers, and that's the same for wildlife.'

By day, terns use Birmingham's canals to migrate and by night Nathusius' pipistrelles and noctule bats navigate along its length. As insect numbers have increased, they've been swiftly followed by fish – local anglers have found perch, roach, rudd, pike and even bullhead in this stretch of canal alone. Bullhead are nocturnal bottom dwellers, with large, flattened heads and toad-like eyes, making them tricky to catch. They require a clean, silt-free stony substrate with well-oxygenated water where the females can build a 'nest' of stones, under which they lay their yellow eggs.

The influx of fish has prompted the arrival of kingfishers, a flash of blue against the grey brickwork, which now feed along the canal all year round.

But what has taken everyone by surprise is the recent appearance of otters all the way into the city centre, something Paul confesses he would never have believed possible just a few years ago. Although they've only been captured on camera a handful of times, the otters can be tracked by their droppings, or spraints, which they deposit under the bridges to demarcate their territory. Despite their deceptively cute appearance, otters are fiercely territorial and can be vicious when it comes to their rivals, so drawing these scatalogical dividing lines is a way to avoid potentially costly confrontation.

Otters need big, connected food corridors, so their territories are expansive. Two such territories have been identified within commuting distance of this stretch of canal – one about 3 miles down from the Sea Life Centre and another 3 miles up on the Soho Loop to the west of the city centre. 'They will take a few fish here, a few there, so there needs to be a lot of resources – otherwise an otter's simply not going to come into that area,' Paul explains, 'so otters are one of the best indicators we've got of a healthy waterway.'

Like the Wandle, Birmingham's canal network is one of the success stories that benefitted from the clean-up initiative in the 1990s. But, decades on, it's still a mixed picture. Large numbers of our waterways are receptacles for raw sewage, discharged through overflows that were intended only as an emergency measure during periods of extreme rainfall. Across the UK, there isn't a single river that the Environment Agency considers clean enough for humans to safely swim in.

At the same time, we've been losing aquatic and wetland habitats in both urban and rural locations as a result of development, intensive agriculture and climate change; over the past century, half a million ponds have disappeared altogether.

The state of our urban blue spaces tells us so much about the state of our urban wildlife. A few years ago, my neighbour Georges dug a small wildlife pond in his garden in Camden. Within six months he saw a dragonfly. After a year he found the first smooth newt, hiding under his hydrangea. Another quickly followed, and then another. Like the wildlife that has flocked to the Wandle, or now roams throughout Birmingham's canal network, the success stories show us what's possible. Everyone knows there's no life without water; what never ceases to amaze me is the speed with which that life can return, populate and repopulate, colonise and recolonise our cities when given half a chance.

A DAY IN THE LIFE

In cultural mythology, water often provides a transition point – into another realm, or another life. Springs emerge as if by magic, ephemeral pools dry up and reappear,

teeming with life, while the organisms within them transform and metamorphose.

Of all aquatic creatures, there is one group of insects
that has caused many a philosopher to ponder on the
transience of life and the fragility of our own existence.
The strange lifecycle of the mayfly renders it a living
paradox: it is at once one of the world's most ancient
creatures and one of its most fleeting – hence its Latin
name, *Ephemeroptera*.

Mayfly fossil records date back over 300 million years,
long before dinosaurs roamed the earth, making them
one of our first winged insects. They're famed for a life
that lasts just a single day, although this is only a half-
truth. Even less true is the commonly held belief that they
emerge solely in May; the name mayfly refers to only one
of many species – the green drake – which appears when
the mayflower, or hawthorn, is in bloom. Other species
pop up throughout the year.

A mayfly begins its life as an egg, sitting on the bottom
of a waterbody. Within a matter of days or weeks,
depending on the species and the water conditions, the
egg will hatch and a brown nymph will emerge, with a
three-pronged tail and external gills protruding from its
abdomen. The nymphs spend various amounts of time
underwater – up to two years in some species. Then, when
the time is right, they'll prepare to emerge with a few trial
runs, swimming up to the surface and then back down,
as if building up the courage for their big reveal. On the
final attempt the back of their thorax will crack open.

The adult mayfly inside is covered in fine hairs, repelling the water and allowing the creature to pull itself out of its old body and on to the water's surface, where it will sit, floating downstream while it pumps up its wings, bracing itself for a brave new world.

It's at this point in their lifecycle that mayflies are at their most vulnerable. Some species will emerge en masse in their tens of thousands, gliding down the river for miles. Fish, larger insects and birds need only open their mouths or beaks for an instant protein hit.

As if one outfit change isn't enough, mayflies are unique amongst insects for going through two adult stages during their brief lives. Assuming they escape the water in one piece, they fly to a nearby patch of vegetation where they'll sit and moult for a second time. Once their transition is complete, they will swarm, mate and summarily die, some within twenty-four hours or less, some a little longer. With the final effort of laying her eggs, which drop to the bottom of the water, the female expires, her wings flat to the surface of the water, where the fish will finish her off. Meanwhile, the male heads back to land for his last rites. Who would be a mayfly?

So mayflies spend most of their existence underwater, and there's no guarantee you'll find adults in May – something I soon learn when I head out on 4 May with invertebrate charity Buglife's Conservation Director, Craig Macadam, in search of them.

I take the train up from London to Edinburgh and then on to Larbert, the town where Craig lives. He picks me up from the station and it's a short drive to the River Carron, a largely urban river that flows through central Scotland. Sand martens, white-bellied with short, forked tails, dart above our heads in search of insects, while a crow sits on a rock preening its feathers. An angler grunts a hello before setting up a few metres away; the Environment Agency keeps this stretch well stocked with trout, Craig informs me, and in June there's a run of migrating salmon and sea trout. At night, Daubenton's bats use the river as a corridor, snatching insects off the surface of the water.

The train line I've just arrived on passes over the top of this section across a large viaduct. It's also crossed by a busy A road, the sound of flowing water accompanied by the steady flow of traffic. 'This river used to supply a lot of mills and a big ironworks,' Craig tells me, 'and it was basically used as a sewer. All the drainage from the town went into it – in fact, a lot still does. It's been cleaned up, but you can still sometimes smell the soapy, Fairy Liquid-type smell of sewage in the water.' He points to an overflow discharging into the river as we speak; it could be diverting surface water from the road, but it's hard to know for sure.

Craig has donned his thigh-high waders in order to collect a sample, while I wait anxiously from the safety of the bank. First he takes out a white tray and fills it with some river water. Then he grabs a big net and wades into the middle of the river. Since it rained quite heavily last night, the water level is up and what are now the shallower parts of the river may until recently have been the beach, so in order to take a representative sample he needs to go deep. He pushes the net to the bottom of the river and holds it there for a short while with the water flowing through, before carefully returning to the bank and emptying its contents into the tray.

We're looking for both mayfly nymphs and for the larvae of other 'riverflies', which will give a good indication as to the health of the water. I peer into the tray and immediately find it full of life; nymphs dart about, a waterlouse – closely related to the woodlouse – scuttles along the bottom of the tray and tiny worms – larvae of the infamous Highland midge – coil and flick their tails. The more I look, the more the tray seems to reveal.

Craig takes out a teaspoon and retrieves a mayfly nymph, which he identifies as an olive upright, common in Scotland and the north of England. The overlapping gills down its sides form a kind of suction cup, which it uses to grip the stones of the riverbed and avoid getting swept away. On its back are little wing buds, indicating that it will soon emerge from the water – within the next couple of weeks, Craig supposes. As nymphs, olive uprights will feed on algae, but during their short adult lives they won't feed at all, so focused will they be on reproducing.

'Ooh, that's nice!' Craig continues, picking out the larvae of a stonefly, known as a yellow Sally, which comes out in June. The nymph looks a bit like an earwig; it's hard to tell which end is its head and which is its tail. This is a deliberate ploy, Craig explains, to confuse predators. The hope is that anything that goes for it will opt for the tail, which it can afford to lose. In fact, something seems to have had a little nibble already, as the tip of its tail is missing.

On a roll now, Craig fishes out the nymph of an iron blue dun – a tiny fly when it emerges, its name helpfully describing its colour. Seven pairs of gills on its sides flap like tiny feathers as it attempts to oxygenate the water in the tray.

Craig continues to reel off names, scooping up each individual with his teaspoon. When these larvae emerge as flies, many will take on new and beautiful hues of green, yellow and blue. At present, though, they are all ochres and browns, blending in with the riverbed. For most people, distinguishing between them would prove an impossible task; but on closer inspection, things that initially appeared the same gradually begin to reveal their differences.

New creatures continue to appear, one of the strangest a cased caddisfly. When Craig first emptied his net into the tray only its long, cylindrical case was visible, with an opening at one end and a smaller breathing hole at the other. The creature inside now pokes out its round, maggot-like head, followed by a couple of pairs of legs.

As adults, caddisflies are moth-like insects with wings covered in tiny hairs instead of scales, which fold back along their bodies when they're not in flight. It's the larvae that build the distinctive cases after which they are named, made from material they find at the bottom of the river and held together with a kind of silk secreted from glands in their mouths. These cases reflect their environment – they use whatever resources are readily available. A French artist named Hubert Duprat kept cased caddisfly larvae in tanks, which he furnished with miniature sticks of gold and pebbles of turquoise and pearls. This resulted in tiny glimmering sculptures, built with an exquisite sense of purpose and a strangeness that no human jeweller could recreate.

The cased caddis in our sample has used pieces of gravel and fragments of red brick melded together like armour to build its home. It skulks across the bottom of the tray, its rather unwieldy case swaying behind it. Craig lifts it out of the water with his spoon and it immediately retreats back inside. Without their armour, cased caddisfly larvae are soft and fleshy – vulnerable to hungry fish or at the mercy of the current.

Craig fishes out another caddisfly larva from the tray. This one is caseless, but an equally enterprising creature; it makes a little net in the water, he explains, which it uses to catch smaller prey. 'They are ingenious,' he adds, 'and they can also be quite territorial. They'll make a little sound in the water to say "This is my space. Back off!"'

We estimate that there are well over 100 creatures in the tray and more than ten different species – all from one swoosh of the net – indicating a fairly healthy river. Craig recalls taking a sample from the same spot fifteen years earlier to present to a local primary school and finding very little activity at all. He got in touch with the Environment Agency and they subsequently discovered there had been a pollution incident from a farm further up, which had wiped out 3–4 miles of life in the river. 'If I hadn't taken that sample, they would never have known,' he notes. For this reason, Craig urges people to get involved in citizen science and take samples like we have today. 'You don't even need a net – you can use a sieve,' he adds; 'the most high-tech bit of kit I use is a pair of wellies.'

Craig returns the contents of the tray to the water (before the larger creatures devour the smaller ones) and we start to look for adult flies, although we're not holding out much hope, since the weather has been unseasonably cold and wet of late even by Scottish standards. We turn over a few pebbles on the beach, where stoneflies often lurk, but alas not today. Next, we scramble up the riverbank, where there's plenty of vegetation (including widespread remnants of Japanese knotweed, which the local authority has been struggling to eliminate). Craig sweeps through with a net, bashing the greenery as he goes to knock off any resting riverflies, but his efforts go unrewarded save for a small aquatic midge.

We return to the car and head towards Carron Dams, a local nature reserve. On our way, Craig tells me about his rather circuitous route to his current role at Buglife. As a child he was fascinated by his local river, where he used to collect shrimps and take them home in jam jars. One day he noticed that some of the shrimps were marked with a little orange dot and appeared to swim differently. The dot, as it turned out, was a parasite which alters the behaviour of the shrimp, causing it to swim to the surface, where it becomes visible to birds. The birds then ingest the shrimp, taking the parasite to its final host.

Far from being deterred by this grim tale, Craig was hooked. Later on, when there was a major pollution incident at the river after a diesel tank burst in one of the quarries, he watched as the wildlife died off and then gradually returned, taking notes and tracing its progress. He ended up studying civil engineering at university, hoping to examine the mechanics of rivers; but when this ceased to be an option he left for a job at Stirling Water Board, where he helped to design sewage works. This led to a role as an environmental policy officer at Scotland Water and finally a return to his first love – invertebrates.

We arrive at Carron Dams, which is located just off the main road and behind Larbert High School, the largest secondary school in Scotland with over 2,000 pupils. The reserve is surrounded by housing and industry, past and present, including a factory making plastic baths. At the heart of the Industrial Revolution in the eighteenth century this whole area used to power the blast furnaces of the nearby Carron Works. By the early nineteenth century it had become the largest ironworks in Europe, but after 223 years of operation the company that owned it became insolvent in 1982. Other than a few remaining bits of old slag from the blast furnace, there is little left to indicate its former use – the whole site has been reclaimed by nature and now forms a wetland, with areas of rich fen and woodland.

We enter along a path through a small wooded area and soon hear the call of a secretive water rail, remarkably like the squealing of a pig, as the fen opens up to our right. Craig is keen to show me a springhead, but to reach it we must jump across the water, which, after a few days of heavy rain, is certainly more than a trickle. I just about accomplish this feat, although not without splashing mud halfway up my legs.

The spring rises from a mound by the fence between the reserve and the school. Tiny flies buzz above the water's surface and we're surrounded by clumps of bluebells, some Spanish, with erect stems and bells either side, some British, with the bells on just one side, causing their characteristic droop. Only British bluebells have that sweet, sought-after scent that eludes many a perfumer. Here, it intermingles with the smell of wild garlic, emanating from clusters of white, star-shaped flowers. On either side of the spring are lush green ferns, patches of spongy moss and dozens of discarded drinks cans, plastic bottles and a broken office chair.

In days gone by, there would have been springs like this popping up all over town, Craig explains, but most have since been built over or piped. They usually remain at a pretty steady temperature and flow, but when large buildings are developed on top the ground is compressed, which can increase the volume of water as it comes out – this spring has likely doubled in size since the High School was first built.

Craig clambers into the spring to collect a sample and

then takes out his teaspoon. He fishes out a tiny fresh-water shrimp – the kind that first sparked his interest in invertebrates. 'It's swimming on its side, which means it's a native species,' he explains; 'non-native shrimps swim upright.' He soon finds one such offender, even smaller in size than the first and slightly orangey in colour. At present these non-natives don't seem to be causing much disruption, but things can seem quite benign for some time before the impact is felt, Craig warns.

Next he picks out a minute snail that could quite easily be mistaken for a piece of gravel. 'It's a New Zealand mud snail,' he informs me confidently; the giveaway is the pointed shell, which differs in shape from most of our native species. The snail is swiftly followed by a flatworm, Latin name *Polycelis felina* on account of its tiny tentacles, which resemble cat ears. We find two caddisflies – one caseless, another with a rectangular case made of twigs and old vegetation – and a stonefly nymph, all of which live only in springheads like this one. 'These are the real specialists,' Craig explains, 'and yet most people would just walk past this spring and not think anything of it.'

Craig turns his attention away from the tray and onto the tree trunk I'm leaning against, where he's spotted an adult stonefly. 'That's really interesting – I've not seen one of those here before – I might keep that and take a look at it,' he says, taking a small plastic pot from his inside pocket, trapping the fly on his first attempt and tucking the pot back into his jacket for safekeeping. Stoneflies will live for up to a month while looking for a mate, so it should survive the trip, he tells me; 'although I do get told off for keeping samples in the freezer.'

Stoneflies aren't particularly good at flying, which makes them relatively easy to catch, but they are strong communicators, engaging in a behaviour known as 'drumming'. While sitting on a hard surface, the male drums a signal. If there's a female nearby, she will respond in kind. The male will move closer and drum again, and so they will continue until they find one another and mate.

Craig still has his eyes fixed on the tree trunk. He takes out another vial and efficiently captures an adult caddisfly. I'm starting to wonder if he has superhuman eyesight, but he claims it's just a matter of tuning in on a micro level. 'When most people walk around, they're probably looking at the big picture,' he explains, 'but I'm looking for movement and I'm focused on the detail.' To prove his point, he hands me both flies in their matching pots so that I can play spot the difference. Both are small and brown, but the caddis has tent-shaped wings with no antennae, while the stonefly has long antennae and wings that lie flat along its body.

As we're about to head back, Craig spots an owl midge, with brown dappled wings and two fuzzy antennae. We make a quick pit stop to a pond thick with rushes, its water silty and black, where we find springtails bouncing across the surface, a green-coloured, non-native shrimp, an aquatic earthworm, some bloodworms – the larvae of non-biting midges, coloured red by a haemoglobin-like substance that helps them survive in stagnant water – and a water beetle larva. The latter is a fearsome predator, with jaws like hypodermic syringes which it uses to inject an enzyme into its prey and then suck out its juices. In adult form, it's capable of taking tadpoles and even small fish.

Craig drops me back at the train station, and just as I'm about to jump out he spots another caddisfly on the car window. 'It must've come off my clothing,' he guesses; 'they're poor fliers, so they just sit on you where they land.' He calmly takes out another glass vial from his jacket (which is beginning to resemble Mary Poppins's handbag in its unexpected and seemingly infinite contents) and catches it against the window, returning the pot to his pocket.

I assume that this is the end of my adventure with riverflies until, sitting on the train home, I recognise one of the stoneflies with the long antennae on the table in front of me. Fortunately, it's dopey enough that even I'm able to catch it with my hands. I release it at the next stop before the train doors close, hoping it will find another spring to fulfil its life's purpose and bequeath the world more stoneflies.

EARTH AND WATER

In my childhood home we had a large pond in our back garden, shaped like a figure 8. The pond was there before we moved in but my parents cleaned it out and re-lined it, adding water lilies and clumps of oxygenating water-weeds, along with goldfish and koi carp.

On one memorable occasion, a huge, ghostly white koi leapt clean out of the water in an ill-thought-out bid for freedom. Witnessing the event, our Beagle Lily sat howling next to the fish as it flapped about on the patio until my mother came out and returned it to the pond.

Fish do not belong in a wildlife pond; plants, insects, frogspawn, tadpoles, other fish – they'll gulp down whatever will fit in their mouths. Only toadspawn is immune, containing the same toxin that adult toads produce in their skin, which creates a foul taste and discourages other creatures from eating them. (Foxes are one of the few animals that have learnt a gruesome trick to get around this, by flipping the toad on its back and ripping out the underside.)

Over time our fish numbers dwindled, thanks to the local heron and a pair of ducks we added to our mena-gerie. But even in the fishes' heyday, our pond boasted a breeding population of frogs, toads and smooth newts, along with an abundance of pond skaters, water boatmen and other invertebrates. I would spend hours gazing into the water, waiting for an air bubble to appear on the sur-face, soon followed by a pair of brown frog eyes, round pupils encircled by pale-yellow rings, or golden toad eyes, with horizontal, slit-shaped pupils, or the dragon-like head of a newt, shimmying up to take a breath.

During early spring, I'd wake to a cacophony of croak-ing frogs, the males puffing out their delicate throats in a bid to attract a mate. This would go on for hours until a suitable female was found to cling on to (or sometimes another perplexed male). There the male would stay until the female laid her eggs, which he'd spray with his sperm to fertilise.

Along with jelly-like clumps of frogspawn, I'd find lines of toadspawn, like strings of beads, and newt eggs, each folded carefully into a leaf. Every year we'd fill plastic ice-cream tubs with a selection and take them into the workroom where my mother made her wooden sculptures. The little black dots would begin to wriggle within their transparent eggs, munching their way out and emerging as tadpoles and newt larvae. As they increased in size, the tadpoles' eyes would grow bulbous and two little buds would appear, first at the back and then at the front, sprouting into tiny legs. The newt larvae, longer, leaner and more fishlike in shape, would take on the form of their parents and lose their external gills, while the tails of the tadpoles would shrivel and drop off. Metamorphosis complete, in a few short weeks we'd be left with perfectly miniaturised versions of the adult creatures which my mother would hastily take to the pond, after one fateful year when she found half a dozen dried-out froglets behind her bandsaw.

It was therefore no coincidence that my first friendship at secondary school was formed with a girl named Frankie – still a close friend now – when our form tutor shrieked in horror at the sight of a 'frog'. 'It's a toad,' Frankie and I chorused in unison, rolling our eyes. (Frankie went on to study Zoology at university and is now an animal behaviourist.)

These days I rarely encounter amphibians – their numbers have declined drastically in both urban and rural locations, where they've been hit from almost every angle. They are prone to various diseases; ecologists are particularly worried about an infectious fungus known as chytridiomycosis (chytrid for short), responsible for what is being described as the sixth mass-extinction. Thought to affect the skin tissue, chytrid results in problems with respiration and water uptake and is already causing the loss of many tropical species of amphibian. It has been found in a handful of locations across the UK, introduced through the importation of non-native species. The fear is that should more non-natives find their way into our waterbodies, it could trigger a similar catastrophe over here.

But for now, the biggest issue facing our amphibians is habitat loss. Without a suitable place to breed, they are forced to migrate across greater distances, which in turn results in high mortality – a staggering 20 tonnes of squashed toads alone are pulled off our roads every year. At the same time, increased pesticide use has hit amphibians' main source of food – invertebrates.

'This is why I'm so glad you're covering amphibians,' says Emily Millhouse, a project manager for amphibian and reptile conservation charity Froglife as she greets me on a warm September afternoon at the Greenwich Peninsula Ecology Park in south-east London. While

186

iconic mammal species like hedgehogs and water voles get a fair bit of attention, amphibians and reptiles are less understood, she argues.

Emily has spent the past four and a half years at Froglife encouraging people to build and restore ponds in the capital, and despite her outdoorsy appearance she is a true urban ecologist. 'I grew up near Bethnal Green [in London's East End] in a high-rise flat,' she tells me, 'and it was very much my mum taking me out to parks to see wildlife that shaped me. I don't think you'd find me out in the countryside – I want to inspire the next generation of kids that don't have a back garden.'

Greenwich Peninsula Ecology Park was part of the development work that took place on this section of the River Thames in the run-up to the Millennium. When work began on the Millennium Dome (now the O2), these four acres of former Industrial Revolution ironworks were put aside, following considerable pressure from the London Wildlife Trust to mitigate the loss of almost 250 acres of wildlife-rich habitat, the aim being to return them to the wetland environment that used to encompass the whole peninsula. The site is surrounded by housing, and local residents and supporters are fighting a constant battle to prevent further high-rise development from enshrouding it in constant shadow.

The park houses two separate freshwater lakes – one outer, one inner, both fed by a large borehole that pumps up from the water table. The interior lake is surrounded by marshland, which hosts snipe and water rail in winter and reed warblers and swifts in the summer. There's also a small patch of alder woodland, a wildflower meadow and a few smaller ponds. Since its creation, the park has become a thriving habitat for amphibians, with an abundance of smooth newts, common frogs and the occasional toad in the compost heaps.

Emily takes me to the wildflower meadow, which currently looks like a patch of old grassland, the flowers being over for the year. She shows me a small pond, partially obscured by the tall grass, which I would probably have otherwise overlooked. The pond is covered by a green metal grate to stop humans from falling in, with gaps big enough for smaller creatures to climb out. In spite of the pond's modest size, both frogs and newts breed here in springtime (although it's not quite deep enough for toads, which need depths of at least 60 centimetres).

There may still be some newt larvae in the pond, Emily says; if they've not reached a sufficient size by autumn, they'll sink down into the silt at the bottom. Suspended in their juvenile form in a process known as neoteny, they will retain their external gills, branching from either side of their heads like furry antlers, and complete their transition as the weather warms up in the spring.

For those that have a garden, a sunken pond like this one is the best way to attract a range of wildlife, Emily explains, but a container pond is fine too – anything bucket-sized and upwards; the important thing is to make sure that creatures can get in and out, through adding some taller plants or creating a ramp using piles of stones or a wooden log. Allowing your pond to fill with rainwater is best, but tap is fine too, so long as it's left for a few days for the chemicals to evaporate. 'Any pond is better than no pond,' she adds.

But today we're not here for pond-dipping; contrary to popular belief, outside of the breeding season most amphibians spend more time on land than in water, so it's under the nearby logs that Emily suggests we start looking. Adult frogs and newts will be after somewhere to hibernate or lie dormant over winter and these log piles, or an adjacent dead hedge made from old branches and leaves, will serve their purposes nicely, with the long grass adding plenty of places to hide. Toads, which are somewhat rotund, need more space to dig down, forming snug little burrows.

Emily carefully lifts a slice of old tree trunk and, sure enough, beneath it we find a cluster of smooth newts in an array of sizes and hues, from sandy-brown to almost black. I've only ever seen newts in the water before, snaking up to the surface or thrashing about like tiny alligators as they hunt for insects, tadpoles and small crustaceans. In terrestrial form they are less elegant, but no less endearing. Although similar in appearance to common lizards, their skin is smooth and matt instead of shiny and scaled, and they have four toes rather than five on their front feet.

The newts' bodies form shapes, resembling letters and punctuation marks; a mass of Cs, Ps, Ss and commas. We count eleven in all – when they find a good spot, they don't mind sharing. Other species including frogs will crowd in too, if there's space.

'Frogs and newts have an interesting relationship,' Emily explains. 'I've found that whenever you build a small pond like this one, the smooth newts move in first. They seem to have the ability to find water in London like nothing else. Then, if frogs move in, the newt population booms, because they'll eat the frogspawn. They're not daft – they know free protein when they see it, and they keep their own eggs well hidden. But over time a balance will come. If there's too many newts they'll go off to conquer other people's ponds and the frogs will rise in number.'

Frogs, newts and dragonflies are likewise engaged in a battle to rule the waters. Dragonfly larvae are fearsome predators; some will spend up to five years terrorising the pond, feasting on frogspawn and newt eggs before they finally leave for good on two pairs of strong, transparent wings. But when the amphibians reach adult size they will soon turn the tables, hoovering up the dragonfly larvae as the hunter becomes the hunted. 'It feels quite fair,' Emily reflects; 'especially when you consider that a frog could be laying more than 2,000 eggs.'

Newts are most active in the evening, just after dusk, staying hidden during the day to avoid predators, of which they have many – kingfishers, herons, foxes, badgers and domestic cats, to name a few. With little in the way of defence, they do have one miraculous skill: they are masters of regeneration. If they lose a limb, they'll regrow it, and they can do this multiple times throughout their lives.

The newts we've unearthed remain perfectly still, as if unaware they've been exposed. But as the sun starts to slowly warm their skin, they begin to wriggle. Emily reaches down and lifts one up to take a closer look. 'The best way to pick up a newt is from the fat part of its tail,' she tells me; 'they do have ribs like us, but they're very brittle, so this way we'll cause no discomfort.' During the mating season, male smooth newts have a wavy crest, which runs from the base of their head all the way down their body. This differs from the mating attire of the otherwise similar male palmate newt, which develops a filament at the tip of its tail. But by this time of year both have lost their crests and the females are always hard to distinguish. Emily, an experienced newt handler, flips the creature over to expose its orange stomach and spotted throat, confirming this is a smooth newt; the throats of palmates are translucent and pink, she explains. She can't be certain of its sex, but the large number of spots and the vibrancy of its belly suggest that it's male.

She transfers it to me, and for a moment it sits on my palm playing dead before squirming off the edge of my hand and, to my relief, landing unscathed on the ground below. ·

Emily puts a small stick down before replacing the log, in case anything has moved, to avoid crushing its inhabitants. We peek under a couple more logs with a 100 per cent success rate, for newts at least. Emily keeps a note of our findings – over forty smooth newts in one afternoon.

We move on, this time in search of frogs, passing an ephemeral pool which has dried up entirely, revealing the shingle beneath. Pausing on the boardwalk by the outer lake, we spot a male common darter dragonfly, deep red in colour with black-veined wings and large, brown compound eyes. Emily shows me a trick she uses with the groups of children she takes out pond-dipping, holding one finger up slightly above the reeds. The dragonfly briefly lands, before realising his mistake and whirring off across the water.

With so much habitat on offer, I marvel that the newts have settled around one tiny pond. But lakes like this are full of would-be predators: ducks that will gulp a newt down whole and fish that will eat their young. The latter appear after people break into the park at night and dump fish fry, with the aim of returning and fishing them out once they've fattened up. To my disbelief, Emily informs me that this is an ongoing issue in sites across London. Removing the fish without causing harm to other creatures is both costly and challenging.

Having so far failed to spot any frogs, we head to a section of path which the local volunteers have been clearing, where we've been told there are dozens of froglets. We tread carefully through the long grass and for a moment I'm sure I've disturbed one, until I realise it's a large cricket. We soon have the volunteers searching too; just as we're heading off one calls us back triumphantly, cupping a delicate young frog in his hands.

'The best way to hold them is by their meaty bits – their thighs,' Emily tells me as she assumes custody of the frog, its heart visibly beating through tissue-paper-thin skin.

The colour of common frogs varies quite dramatically, from greens and browns through to oranges and reds. This one is a golden ochre on top and soft limey-green underneath, with pinkish fingers on each of its feet. It has a characteristic dark eye mask and stripy back legs which, when outstretched, are longer than the rest of its body. It's about a year old, Emily guesses; a more seasoned frog would be croaking at us – when cornered they soon start protesting.

She holds the frog into position and takes a snap on her phone for the Froglife Instagram page, its wide, iridescent eyes glittering in the warm autumn light. 'I want that classic smiley frog-face shot,' she jokes. Photo shoot complete, she deposits it onto a pile of old vegetation, where it pauses for a moment before taking an almighty leap, disappearing into the leaf litter.

A WATERY WASTELAND

In July 2020, Prime Minister Boris Johnson gave a speech on job creation, in which he urged the nation to 'build, build, build'. And in so doing, he waged war on one of our rarest amphibians: 'Newt-counting delays are a massive drag on the prosperity of this country,' he blustered, referring to the great crested newt, which is a protected species under British law.

For development to take place on a site occupied by great crested newts, adequate mitigation must be provided and the newts translocated to a suitable habitat elsewhere. Anyone found guilty of disturbing their resting places, breeding sites, or taking their eggs is liable for an unlimited fine and up to six months in prison.

Johnson's comments soon faced wide condemnation, with the head of the Wildlife Trusts, Craig Bennett, describing the speech as 'pure fiction'. 'It may sound funny referring to newts,' he continued, 'but actually it was rather sinister. In the environmental movement we know referring to newts is a dog whistle to people on the right of his party who want environmental protections watered down.'

Great crested newts, or GCNs as they are affectionately known amongst conservationists, are increasingly scarce across their native range. In fact, the UK is their last stronghold in Europe, and therefore the world, and yet here too their numbers are declining rapidly, mainly due to habitat loss. This is exacerbated by the fact that GCNs are a little fussier than other amphibians when choosing a suitable home. Their preference seems to be for shadier ponds, but they will select some waterbodies while rejecting others for reasons that aren't always apparent.

Another issue is hybridisation with other newt species, such as the alpine or Italian crested, which has been happening across mainland Europe. Alpine and Italian crested newts aren't found here unless introduced, and due to its island geography the UK is one of the few places where great crested newts are retaining their singularity as a species.

While they are rare, it's possible to find great crested newts in urban locations, especially the outskirts of cities. In London, for instance, there are populations in Waltham Forest, Epping Forest and Richmond Park. There's also a thriving band of GCNs at Spitalfields City Farm in metropolitan Tower Hamlets, where a 2-metre mural was unveiled in their honour. The newts' presence in and around the farm's ponds is a bit of a mystery – it's thought they may have come from a nearby garden where they were introduced, but they've happily lived at the farm for decades.

However, it's not London but the southern outskirts of Peterborough where my dad (who has a particular fondness for newts) and I head out on a drizzly day in early May in search of these elusive amphibians. After a short train journey and an even shorter cab ride we arrive at Froglife's largest nature reserve, Hampton, to the surprise of our taxi driver, who asks us multiple times whether we're sure this is where we want to be dropped off.

As a Special Area of Conservation (SAC), SSSI and Natura 2000 protected area, Hampton is closed to the general public except during organised events. But it's open to a dedicated team of volunteers, whom we will be joining today as they conduct their bi-weekly reptile survey.

The 300-acre site is made up of post-industrial wasteland. As a former brickworks, its bricks were used to rebuild London and other cities after the Second World

War. The curiously undulating landscape was formed by a series of linear trenches – gouged out by the machines used to extract the clay – and the resulting spoil heaps, which create a network of mounds. Over time, nature reclaimed the site; the steep humps of spoil were overtaken by grasses and scrubby vegetation and the trenches filled with rainwater, sealed in by the clay beneath to form 320 ponds.

It is this habitat that attracts and supports the highest density of great crested newts in Europe, and indeed the world – tens of thousands of them. And the newts aren't the only notable residents: Hampton is also home to slow-worms, grass snakes, dragonflies, butterflies, water voles, migratory birds and adders (the team are careful not to advertise the latter too widely due to the risk of persecution, despite the snakes being largely harmless). The site is also one of the best places in Cambridgeshire for butterflies, particularly the early-emerging species, including the vibrant green hairstreak, the more muted dingy skipper and the black and white chequered grizzled skipper. Surprisingly, there are very few toads and no frogs at all, although I suppose that, given the army of newts that would merrily devour their young, the frogs are wise to steer clear.

Previously, much of Peterborough was surrounded by land like this, but over time it has been levelled out and developed. On the right-hand side the reserve is flanked by factories and warehouses, owned by vast delivery companies, which have sprung up in the past few years replacing a landscape of open fields. To the left is the large Hampton housing estate; rows upon rows of neat new-build houses, none of which existed a couple of decades ago. The original plan was to continue building, but the land that forms the reserve was saved from development due to the presence of the great crested newts; between 30,000 and 40,000 were translocated from the development site to the reserve.

Between the housing estate and the reserve runs a recently tarmacked road, named – without a hint of irony – 'Nature's Way'. This road will eventually connect with the large village of Yaxley, near a junction of the A1, which will increase the flow of traffic dramatically and the reserve will become an island surrounded by concrete. A wall has already been constructed around the site to keep the newts in, although in reality newts are regularly found beyond its confines in a nearby ditch and in local gardens.

We're met by Ross Edgar, the Conservation Officer who manages the site. Ross used to install CCTV for a living before looking for a career change. He completed a degree in Conservation a decade ago and has been working at Hampton for four years. The site itself doesn't need too much management, he explains – the ponds are deep in the middle, so they rarely dry out – it's mainly a case of cutting back the reeds every now and again. What takes more management, I imagine, are the many eager volunteers, whom Ross is clearly adept at organising.

As the volunteers arrive, they seem prepared for a much more rugged experience than I had anticipated – some sport camo gear and others look like they're about to climb Ben Nevis. Most are of retirement age, although not all – one shy young man tells me he's been volunteering for the past few years in the hope it might lead to a career in conservation. Meanwhile, one of the most senior volunteers, an elderly woman named Judith who conducts the weekly butterfly surveys, is clearly the matriarch of the group, marching around the site with a pair of walking poles. 'You could write an entire book just about Judith,' another volunteer jokes.

In the past, Froglife volunteers have surveyed every pond on the site. They found some to contain smooth newts, some GCNs and some both. These types of survey are conducted at night – the best time to see newts in the water – by shining a torch into the ponds and scouring the margins. But scrambling up and down the steep slopes in the pitch-black with 3 metres of water beneath is a big undertaking. Today we have a much simpler task: to search for newts and reptiles underneath 150 pieces of corrugated metal and carpet tiles that have been strategically placed around the reserve. Both the rectangles of metal and the square tiles heat up in the sun, attracting cold-blooded creatures in need of a warm place to hide during daylight hours.

Since the weather has been unseasonably cold, the volunteers aren't feeling particularly optimistic, but my dad and I remain hopeful. Ross assigns groups and gives each of us a section to cover. We are told to join Anne, a retired supermarket worker who has a kind but cautious energy and an impressive knowledge of the site. We're also accompanied by a younger woman called Emily, who used to work for Froglife before recently taking a job at London's Natural History Museum (not to be confused with the Emily still working at Froglife, who showed me around the Greenwich Peninsula Ecology Park).

Anne warns us to be careful, as the ground is uneven. We glance down and see flecks of old brick beneath our feet and an abundance of brown-lipped snails, their shells a host of sweetshop colours, from yellow sherbet to candied orange. The reserve is split into sections marked by letters: we've been allocated the Bs and Cs, and fortunately for us Anne can remember the exact locations in which to find the metal tins and accompanying mats. Reptiles prefer the metal, she tells us, while newts will opt for the cooler, damper environment of the mats. We take it in turns to carefully lift each numbered square. On the underside of each mat we uncover a sea of ants, which tumble down and make frantic attempts to protect the colony's eggs.

After making our way through most of the Bs we've still yet to find a great crested newt. We pass through an area that was recently set ablaze by teenagers lighting a huge bonfire. Just a few years later it has recovered completely, the tall grasses returning and reed maces crowding the ponds. This wasn't the first time trespassers found their way on to the reserve, nor is it likely to be the last, with urban life all around; Anne often spots people walking their dogs or using the slopes as makeshift jumps for their bikes.

As the weather continues to vacillate between drizzle, wind and glimmers of sunshine, we disturb a small muntjac deer on one of the paths which quickly scarpers. Anne draws our attention to the loud song of a Cetti's warbler, one of our most recent colonists, which only began breeding in Britain in the 1970s. These birds now overwinter on the reserve, she tells us.

I'm close to giving up hope as Emily lifts another mat in an unremarkable spot, shaded by a patch of scrub. Beneath it are two dark creatures, curled up like a pair of black olives, their moist skin glistening. Great crested newts are about twice the size of smooth and palmate – our other two newt species – but judging by their size this pair are probably last year's young. They're black all over, including their eyes, with their toes the only exception – the four little blobs on each foot are an orangey-brown. As juveniles it's hard to determine their sex – they lack the jaggedy crest developed by males as breeding attire, which makes them look like tiny dinosaurs in the water.

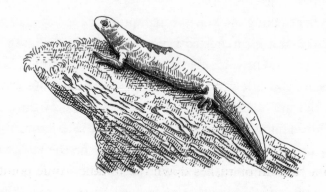

GCNs breed between March and June, during which time the males will woo the females with an extravagant courtship display, performing an aquatic handstand as they arch their backs and sway their hips and tails, as if under the influence of something other than hormones alone. But for this to take place, the water needs to reach at least 5 degrees, and following the coldest April on record their breeding may be delayed this year, Anne suggests.

The newts won't reach sexual maturity until they are between two and four years of age, which is less surprising within the context of their lifespan as a whole – the grand old dames of the newt world, GCN females can live for up to twenty-two years, nearly twice that of smooth and palmate newts. Some of the newts on this site must have experienced significant change in their own lifetimes as the habitat around them has shrunk through continuous development.

After the first pair of GCNs we soon find another juvenile – this time under a warmer mat in a patch of direct sunlight. Its skin is covered in tiny bumps and I notice not a crest but a thin stripe, starting just behind its eyes and running all the way to the tip of its tail. Its toes are especially orange, as if painted with brightly coloured nail polish.

We carefully replace the mat and pause by the bank of one of the larger ponds, surrounded by dense foliage, with a column of rushes down the middle. Anne points to a badger sett at the top of a steep bank on the other side. Amidst the smell of damp vegetation an unexpected scent wafts into our nostrils – Emily describes it as beans on toast; to me it smells more like crumpets. Anne quickly unravels the mystery: one of the new factories behind the reserve is an industrial-sized bakery.

We pass a cluster of large ant hills, which resemble the landscape itself in miniature, and I flip one more mat with a few rabbit droppings sitting on top. Underneath are four more athletic GCNs. Upon finding itself exposed, a larger adult quickly wriggles down a crack in the ground, flashing us with its orange belly covered in black spots, which form a pattern as unique to each individual as a fingerprint. The remaining three newts are juveniles – the second quickly follows the adult while the third tries to hide under the fourth, which clambers down into the crack too; but, space now being at a premium, its tail is left poking out.

Returning the mat to its original position, we head back towards the centre of the site to reconvene with the other volunteers. On the way Emily stops in her tracks, spotting a smooth newt on the path in front of us and moving it to a place of safety in the long grass. We soon reach the rest of the group, where we begin to discuss what we've seen. Most people have similar findings, although one retired couple were surprised to discover a small plastic bag filled with white powder under one of the numbered metal tins, and an animated discussion takes place as to whether it should be removed or left alone in case the owner returns.

The powder is soon forgotten when Ross reveals he

caught sight of an adder close to the car park and we all decide to traipse in that direction in the hope that she's still there. Ross leads us up towards the summit of one of the spoil heaps where, sure enough, the adder is sunning herself on top of one of the pieces of metal. It's the first time I've seen a wild snake in this country and I'm struck by how small, short and chubby she is, her scales gleaming in the early afternoon light.

Adders are shy snakes and she soon senses our presence and slips off into the long grass. Meanwhile, I'm reminded that in safeguarding a species like the great crested newt we protect a multitude of others – just like this adder – by retaining these pockets of habitat within our urban areas.

OUT OF THE BLUE

Looking out across Hogganfield Loch in the north-east of Glasgow, I can see plenty of prime water vole territory. There's long grass and aquatic plants for the water voles to eat, and ditches with steep sides ideal for constructing their burrows, where they can add holes both above and below the waterline, the latter forming an escape route from predators as they shoot into the water with a characteristic plop.

Hogganfield gets its name from the castrated rams, known as hogs or hoggets, that grazed here in the seventeenth and eighteenth centuries before meeting their fate at the slaughterhouse. It is one of many naturally occurring kettle ponds, formed by glacial slip during the last Ice Age, but over the years it has been subject to considerable human intervention. Up until the 1950s the surrounding area was mostly marsh and farmland, but since then large-scale housing development has engulfed most of the land around the loch.

More recently, work has been carried out to soften and re-naturalise the loch edges through the creation of shelves, carpeted with lush vegetation, surrounded by patches of marshland and wildflowers. Hogganfield Park is now a designated Local Nature Reserve, connected to Cardowan Moss Local Reserve and other sites across Glasgow and beyond, creating a network of wetland, woodland and grassland habitats which form the Seven Lochs Wetland Park.

As the sun reflects off the water, dozens of swans prepare for the day ahead, arching their long necks, preening their feathers and scooping and filtering the water with their orange bills. The loch is still a popular location for birdwatchers, well known for its overwintering migrants. Whooper swans often arrive from Iceland and recently a spoonbill has been spotted on the island in the middle, where a large heronry is located. An osprey has also been filmed fishing in the loch.

Water voles used to be found in waterbodies like this, and along rivers, streams and canals across every corner of Britain, but they've since suffered a long-term decline. Now absent from over 90 per cent of their former strongholds, they have legal protection across the UK. Predation by non-native American mink has had a worrying impact, resulting in local extinctions in some areas. Poor water quality through direct pollution or eutrophication has stifled their food plants. But it's the all-too-familiar story of habitat degradation and fragmentation that has caused the biggest issues, as we've straightened, drained and culverted the blue spaces that used to provide and connect their habitat.

Although water voles do live at Hogganfield Loch, they can be quite a challenge to find, and Cath Scott, a Natural Environment Officer at Glasgow City Council, is keen to show me some of the more 'unusual' places in which Glasgow's water voles can be found. For these reasons, we're not even going to try our luck here – Cath simply suggested the loch as a convenient meeting point that's a ten-minute cab ride from the city centre.

We're joined by Dom McCafferty, a Senior Lecturer at the University of Glasgow's Institute of Biodiversity, Animal Health and Comparative Medicine. Dom and Cath have been working together for the past few years to learn more about Glasgow's water voles, mapping their distribution and habitat features. This work was sparked by a surprise discovery in 2008, when Glasgow was found to have a booming water vole population; it's just that most of them weren't living anywhere near the water.

Since then, water voles have been popping up from little holes in parks and other areas of urban grassland across the north-east of Glasgow. 'Glasgow has always been considered a stronghold for aquatic water voles,' Cath explains, 'and we knew we had traditional water voles living along our watercourses, in places like Hogganfield, and then up in the north-west of the city and down in the south. And we noticed they were moving towards marshes as well as ditches and burns. Then eventually we found the grassland ones in the north-east of the city, and wondered what on earth they were doing there!'

Glasgow is by no means the only city in the UK to report sightings of water voles, but it is the only place where those sightings are far removed from the nearest expanse of water. Until recently, British water voles were believed to live only in riparian habitats – along the water margins and banks. 'There's 227 species of aquatic plant they've been recorded eating,' Cath tells me, 'and most of our grasslands have only five to choose from. And yet here they are. Once you're aware of the grassland water voles you start to find more and more.'

As Cath kept discovering water voles, she soon became the go-to person for anything concerning them. 'My background isn't even in zoology,' she confesses; 'I'm a botanist. I don't think I'd even seen a water vole before. But botany includes studying plants and habitats, and protecting habitats is key to conserving a range of wild-life.' Her role previously involved creating wildflower meadows across the city, but when the voles moved in, due to their protected status she was no longer able to manage the meadows, 'So I ended up looking after the voles instead!' she quips.

Elsewhere in Europe, water voles are sometimes found

in dry grassland habitats, so the situation isn't entirely without precedent, and their only real adaptation to an aquatic lifestyle is the ability to close their ears to keep the water out. But seeing as they've never in the course of British history been recorded more than a few metres away from the water, it's still pretty peculiar. 'Aye, it's a bit of a mystery as to precisely why they're doing this,' Cath laughs, 'although we've got various theories . . .'

One of those theories is that the water voles may have moved to avoid predators, especially the invasive mink. Female mink can just about squeeze into a water vole burrow, making them particularly deadly, but they prefer watercourses to grassland habitats and tend to keep away from built-up areas. There are, however, few reports of mink in and around Glasgow.

Another theory is that the water voles were forced to seek new pastures as Glasgow's blue spaces began to vanish. During the creation of the M8 – Scotland's busiest motorway, which runs between Glasgow and Edinburgh – a section of canal was drained entirely. It was nearby that the first population of land-based water voles were found, when the council's pest control officers were called out to investigate a 'rat infestation' in an area of scrub between the motorway and a row of small terraced houses. Meanwhile, cuts in council budgets led to reduced cutting of grass, and in areas of greater deprivation in the north-east of the city this ended up providing food and cover for the non-water voles.

Could it be that through Glasgow's terrestrial water

voles we can trace the ghosts of the city's lost rivers and waterways?

The one complicating factor is timing; the motorway was built in the 1960s, over forty years before the first grassland water voles were discovered. But perhaps the initial numbers of voles were small, with the population barely hanging on after the canal disappeared, until for some unknown reason things began to pick up in the early 2000s. 'Once a population starts to establish itself and ends up with more breeding individuals, you get larger groups dispersing where there are corridors, and this can result in a boom,' Dom explains.

Dom and Cath have recently discovered that, in the absence of a watercourse, the water voles are using the M8 as just such a corridor, along with other animals including deer. Odd as this may seem, the motorway does have its benefits, providing a continuous strip of grass along its length with few predators. The voles must have become accustomed to the roar of traffic and may even attempt the perilous journey from one side to the other. 'There have been sightings of small things crossing the motorway,' Cath tells me, 'although we don't know how many of them make it.'

When it comes to Glasgow's water voles, it seems we have more questions than answers, but I'm still anxious to see some of these creatures for myself, so we head off. Dom and I jump into a university van that he's borrowed for the day and Cath leads the way in her small car. We soon arrive at Cranhill, a large residential area consisting of a patch of grassland, now Cranhill Park, surrounded by housing schemes, which boasts the highest density of water voles in the country – 156 per hectare. (In reed beds the number is generally between forty and fifty.)

Before travelling up to Scotland I googled Cranhill, expecting to read about its water voles. Instead, I encountered plenty of articles in the local press detailing a number of grisly murders. Cranhill was first developed using public funding in the early 1950s and composed mainly of four-storey tenement blocks. Over the following decades, three colossal grey high-rises were added, along with rows of terraced maisonettes. Cranhill quickly became known for being one of the most deprived areas in the UK – in the 1990s and early 2000s, the media with its usual sensitivity labelled it 'Smack City'. Today, the council is still working on local regeneration, with the creation of new affordable housing.

We park up outside the tower blocks and head into Cranhill Park, which is in fact another spot with a missing waterway – a burn that has been culverted. It does emerge as a trickle in one part of the park, but it has been thoroughly concreted, making it wholly unsuitable for water voles.

The first thing I'm struck by is how unremarkable the park appears – grass, path, scrub, a few trees and about a dozen jackdaws foraging for insects. Upon closer inspection, I can see little mounds of earth where the ground has been excavated, appearing every few metres or so. Without the presence of Cath and Dom I would have assumed these mounds were molehills, although they are a little on the small side. In some cases, water voles have been known to actually take over where moles have left off, or even cohabit with them, but there aren't any moles in this park, Cath informs me.

The piles of earth are called 'tumuli', Dom explains, and next to each one is a small oval-shaped hole – an entrance into the water voles' burrow systems. Water voles produce these tumuli in wetland habitats too; it's just much harder to see them since they get lost in the water. 'Ordinarily you're searching for ages to find them, but here you're tripping over them,' Cath jokes.

Although water voles generally display a preference for gentle slopes, Cranhill Park provides clear evidence that they'll also burrow down into flat land when space is at a premium. There seems to be little that deters them, with as many burrows next to the path as there are elsewhere. Perhaps, Cath suggests, they actually choose to be closer to people because human footsteps scare off other predators. Without any water to retreat to, their options are limited – race back to their burrow or bite. Water voles use their teeth for digging so they are surprisingly strong and sharp, but retreating is surely a better bet when faced with the likes of foxes, buzzards, herons and gulls.

Dom and Cath's team have worked with specialists to survey the water voles' burrow systems, between 40 and 60 centimetres below the surface, using a ground-penetrating radar. This has revealed what resembles a vast road network, with interconnected chambers of various sizes for rearing young or storing food for the winter. The voles are active all year round, but they spend more time underground in the colder months, munching through their larder or nibbling on roots below the surface. 'Apparently, they overwinter in wee family groups,' Cath tells me, 'which sounds particularly cute.' She pauses. 'I'm supposed to be professional here. Not at all cute,' she laughs.

Water voles can live up to the age of three, but most don't make it beyond twelve months and up to 60 per cent will die over winter. They maintain their numbers through breeding; the females give birth to three or four litters of five to eight babies a year, beginning in March if it's warm enough and continuing through until early autumn. Water vole babies are born blind, hairless and helpless, but they quickly become independent, weaning at just fourteen days.

In a conventional watercourse habitat the females will mark out their territory during the breeding season with piles of droppings. But in these grassland habitats, Cath tells me, you won't see the latrines until later in the year, or sometimes not at all – perhaps they've been moved underground, or as a result of close-knit urban living the water voles are forced to become less territorial. Either way, it is important, because ecologists generally use the latrines to confirm a breeding population which demands protection, as opposed to a few isolated individuals.

All this breeding requires a lot of energy, and the water voles need to eat 80 per cent of their bodyweight every day just to survive. Research by an ecologist named Robyn Stewart, who has been working with Cath and Dom, shows that the voles are active throughout the day, eating and sleeping on four-hourly rotations. Early May is when water voles are not only at their most active, but also their most visible, the grass still being relatively short. While this makes it the best time of year to see them, it's a risky period for the voles. In summer, when the vegetation grows high, they'll have much better cover, tunnelling through the long grass.

Cath tells me to 'keep an eye out for a wee head popping up'. All the signs' are in our favour – there are plenty of fresh piles of earth, indicating recent excavation, and patches of dead grass where the voles have been feeding on the roots from under the ground. We find little clumps of dried grass too outside the burrow entrances, which they use as bedding, suggesting that they have been busy spring-cleaning. But despite all this evidence, half an hour passes and we've still yet to spot a single water vole.

Cath wonders if we're standing too close to the burrows, or perhaps the jackdaws have spooked them. We try hiding behind a tree just off the path instead, keeping very still. Another fifteen minutes pass.

Growing restless, Cath directs us to the other side of the park, where last spring the council built a sustainable urban drainage system as part of a Glasgow-wide green infrastructure project. The primary focus was to reduce the risk of flooding, but the council was also hoping to create green (and blue) corridors in the process. The solution that was devised for Cranhill Park is pretty ingenious: through controlling the surface water, which now flows into the culverted burn, they've managed to create a small basin. A few water voles had to be relocated in the process, but Cath hopes it will be worth it in the long run.

Surrounded by oxeye daisies, the basin is already thick with reeds and clumps of lemon-curd-yellow marsh marigolds. Three abandoned shopping trollies lie at the edge of the water and a grey wagtail perches on a rock, picking insects off the surface. The water level in the basin will be variable, Cath explains, flooding up to a certain point and allowing the water voles to build tunnels down into the water, as well as higher up on the bank to keep their breeding chambers dry. 'It'll be very interesting to see if they start using it,' she adds, 'because we did something similar at another park where a burn was day-lighted and the water voles moved straight in.'

While Cath is still talking, Dom is peering downwards. He's spotted a hole in the bank, and then another. It turns out the voles are already here.

As we head back towards the path, we bump into a local resident named Rab out walking his dogs. Coincidentally, Rab was the first person to identify the water voles at Cranhill, having lived in the area his whole life. He used to keep pigeons close to the park and some of his neighbours began to complain that his hobby was attracting rats. 'I told them, it's nae rats,' he informs me; 'I used to sit and feed them apples. Rats don't eat apples.' This isn't entirely true – rats are reasonably fond of apples – but Rab was quite right about the water voles.

Since then, much public-engagement work has taken place to raise awareness of the water voles and Rab is clearly proud of the part he played in their discovery.

Cath and Rab continue chatting, about Rab's dogs, the water voles, the local buzzard – 'the buzzard's going through a right bad moult', Rab is saying (I can only understand about half of what he says through his thick Glaswegian accent). Meanwhile, I spot a heron as it lands purposefully on a gentle slope between a few scattered trees. Every now and again it inches slowly forwards, neck arched, head pointed downwards, before pausing, eerily still. Presumably, the water voles can feel a vibration when larger creatures walk over their burrows; hence why it's being so careful.

Within the space of a split second, the heron thrusts its beak down a hole. We all gasp, but its attempt is unsuccessful. 'It saw one and went for it, aye,' Cath asserts.

Herons are known for eating almost anything – fish, rats, voles, moles, snakes, frogs, ducklings; at Camden Lock, near where I live, there's one that thrives on pizza, falafel, onion bhaji and macaroni cheese, furnished by the local street food stalls. This Glasgow heron is in good condition – clearly well fed. 'It definitely knows what it's doing,' Dom observes ominously.

'I don't know if this is something we want to witness,' Cath adds anxiously.

A magpie lands a few metres away and the heron seems

drawn towards it. Now we're all watching intently. 'It's almost as if there's some communication between them,' Dom suggests. Rab tells us he's seen gangs of magpies and gulls mobbing herons in the park for a share of their catch.

A few moments later, the heron coils its neck before striking again with its dagger-like bill. 'It's got one!' Rab exclaims. 'That magpie will alert everything and you'll see more of them soon.'

Dom passes me his binoculars and there it is – the only water vole we've seen so far. Almost black in colour, its back end is hanging out of the heron's gaping beak, tail still twitching. Herons are large birds, but water voles aren't insubstantial – though shorter in length than rats, they're broader in the body and only weigh marginally less. The idea that the heron could swallow it whole seems a bit of a stretch (metaphorically and literally), and yet that's exactly what it does. The vole slowly disappears, a bulge passing down the heron's slender neck, resembling an egg being swallowed by a snake, until only the tail is left dangling from the heron's beak like a piece of spaghetti. 'There's surely going to be some indigestion going on there,' Cath observes, with a slight hint of malice. 'That's real urban wildlife for you,' Dom replies.

Once we're done trying to process the trauma of what we've just witnessed, Cath suggests we make a move, anxious to see some water voles in happier circumstances. (I put to the back of my mind questions of whether the heron, now preening its feathers dispassionately, has left a family of tiny orphans in its wake.)

We leave the park and as Cath and I continue chatting Dom raises his binoculars towards a grassy mound in front of the three tower blocks, on the opposite side of the road from where we're parked. He directs us to look at a spot in the centre of the mound, next to a patch of dandelions. The grass here is shorter, making the little chubby brown head easy to identify as it pops out of its burrow. Water voles found in the north of Britain are usually black, while down south they're mostly brown, but Glasgow seems to be blessed with both. Emerging fully from the hole, it scuttles about, hugging the ground in a movement quite different from the bounding of a rat, for which voles are often mistaken.

Pausing for a moment, back arched, propped up on its hind legs, it peers around cautiously, the high-rises looming behind it, before bobbing back down underground.

With that, we make a move towards our next stop, Sandaig Park, a yet more improbable habitat. Sandaig is a significantly smaller park than Cranhill, flanked by a primary school on the right, an overcrowded cemetery on the left and a busy road with a row of shops in front. The park itself consists mainly of grass, although a few saplings have been planted alongside the school fence, and there's a worn-looking outdoor gym, a children's play area and a couple of goalposts. A few locals gather around a bench with their dogs.

The grass is full of dandelions and clearly hasn't been cut for a while. 'I actually have to tell some of my colleagues, you really need to keep it short, which is contrary to standard biodiversity advice,' Cath confesses; 'otherwise the water voles will take over the whole park and then they'll never be able to cut it again and there will be no space left for local residents!' This site really exemplifies the challenges presented by the water voles' unusual new choice of habitat, she explains. To dig or build near water voles, you have to consult with the relevant authorities and mitigation must be put in place. To cut down a tree or lay pipes, you need a licence. Technically, you can't even mow within 10 metres of where the voles are active. The trouble is, the minute the council stops mowing, the water voles spread even further, forcing the council to retreat by another 10 metres. Next thing they know, the water voles have taken over the whole site, which explains why, when Cath sees a hole next to one of the goalposts on the Sandaig Park football pitch, her heart begins to sink.

'Even when a developer builds houses, you have to decide, do you want the gardens to be wildlife-friendly or not?' Cath explains. 'Of course, normally I'd be recommending holes in fences and hedgehog highways, but in a water vole area, the gardens will end up full of water voles, which is not what everyone wants!'

Some people are happy to share their gardens with these endearing little rodents – Cath shows me a video of a water vole that lives under someone's decking, where it paddles around a water bowl and enjoys a range of fruits left out for it. But others aren't so keen – especially when the creatures start to impinge on what they're permitted to do in their own backyard (Glasgow Council and Nature-Scot, formerly Scottish Natural Heritage, have agreed that telling people they can't mow their own lawns feels like a step too far, and evidence suggests that Glasgow's water voles are pretty tolerant to some disturbance).

'People think human and wildlife conflicts occur in the Serengeti,' adds Dom, 'where you've got lions coming into people's villages. But here you have a real conservation challenge too: a highly endangered mammal that is obviously thriving, and on a national scale you want to be doing everything possible to enhance the population. But on a daily basis, Cath has to solve the problem of how they can coexist with people. And we need to find a solution for that too.'

Dom and Cath hope that the work they've been doing will help, feeding into a new water vole conservation plan for the city. One of the things they're hoping to address is the way in which the water voles disperse. Ordinarily, the voles follow the direction of a watercourse, but in a grassland environment their movements are much harder to predict. Through habitat-mapping and trials in grassland and vegetation management, the team hope to learn more about how to keep the populations connected. The council is already trying to bring back and restore those very habitats that the water voles have been forced to leave behind. Every indication so far is that they are ready and willing to return to the water, where water of suitable quality is made available. Providing and maintaining both blue and green corridors will support the population as a whole by allowing individuals to disperse from the dense, isolated pockets, to the benefit of both water voles and humans. Or that's the theory, at least.

On our way out of the park, Dom raises his binoculars once again, this time having spotted first one, then two, then three water voles in the grass beside the primary school. The first, also the darkest in colour, appears to be digging. The second scurries back towards its hole. The third, a rich chestnut-brown, is probably only five metres away. I watch as it noses about, its snout shorter than a rat or mouse's, its face chubbier, guinea-pig like. It munches on a blade of grass, clamped in place by its tiny hands, unfazed by the stampede of children on their lunch break on the other side of the fence. After a good few minutes it detaches a dandelion stem with its teeth and disappears with it into its burrow.

CHAPTER 5

Look Down: The Underworld

Deep was the cave; and, downward as it went
From the wide mouth, a rocky rough descent;
And here th' access a gloomy grove defends,
And there th' unnavigable lake extends,
O'er whose unhappy waters, void of light,
No bird presumes to steer his airy flight

Virgil, *The Aeneid*

In Book VI of *The Aeneid*, Virgil describes another realm beneath our own as the Trojan hero Aeneas travels deep into the underworld to meet with his dead father Anchises. Descending through the mouth of a cave, he enters the hellish gateway between life and death, known as Avernus, meaning 'without birds'.

This isn't purely the stuff of myth and legend – Avernus was the ancient name for a volcanic crater and the lake within it at Cuma, 16 miles west of Naples. It was said that any bird flying over the lake would fall dead; probably due to the toxic fumes the crater released into the atmosphere. During the civil war between Antony and Octavian, the heir to Caesar, the Roman general Agrippa tried to turn the lake into a military port, and the remains can still be seen beneath the water's surface.

In our modern cities, too, there is a world lying under our feet. Creatures hide in piles of leaves, beneath logs and stones and items of rubbish. Just below the surface, roots penetrate and animals delve. The soil itself is teeming with life; one-quarter of all known species live within it, and a teaspoon of earth can contain more organisms than there are people living on our planet. This includes a whole host of invertebrates such as beetles, centipedes, molluscs and the burrowing earthworms that create tunnels, allowing water infiltration and the root growth of plants, as well as incorporating organic matter into the soil and promoting decomposition through their feeding habits.

But also, and perhaps lesser known, are the microscopic organisms that each undertake a vital role. Bacteria that can convert organic matter into forms that are useful to plants and work in partnership with those plants to break down pollutants. Fungi, which bind soil particles together, recycle nutrients and even suppress diseases. Single-celled organisms called protozoa, which consume bacteria, releasing excess nitrogen that is then made available to the network of plant roots around them. Nematodes, microscopic worm-like creatures that live in the thin film of moisture surrounding each soil particle, aiding the decomposition process and recycling nutrients. And tardigrades – tiny, out-of-this-world-looking creatures of 1 millimetre in length – which feed on plant cells and even more minute invertebrates; sometimes known as water bears or moss piglets, to my eyes they resemble the bag that goes inside our vacuum cleaner. Tardigrades have been shown to survive conditions unheard of in any other living creature. Scientists have heated them up to temperatures of over 150 degrees

Celsius, frozen them down to -272.8 degrees, blasted them out to space – and they've still lived to tell the tale.

As we dig deeper, foundations, basements and labyrinthine tunnels form subterranean caverns through which trains thunder, sewage stagnates and hidden rivers flow as the earth vibrates with human activity. The London Underground – nicknamed the Tube, which is also pleasingly literal – was the world's first underground passenger railway, first opened in 1863. With more than 100 miles of underground tracks it provides a unique environment not only for humans, but for a few other brave species too; even birds – pigeons are occasionally witnessed using the Tube to commute from station to station. After hoovering up any crumbs on the platform and minding the gap, they hop onto the train and amble about the carriage, before heading towards the doors as the train slows down and exiting when they open.

Some creatures are pulled downwards by an irresistible biological impulse. Others end up there unwittingly. Some even adapt to this new environment, a product of urban evolution, living out their whole lives hidden away from the light of day. But unlike the ghostly underworld of myth and legend, the world beneath our cities is very much alive.

SCRATCHING THE SURFACE

It's an unseasonably warm afternoon in late September when I arrive at Norbiton, in London's south-western

outskirts, for a second time to meet with ecologist Alison
Fure. 'I might stink of honey because I've just cleaned out
my beehive,' she warns me.

Today we're not in search of Alison's speciality, bats,
but something of a different class altogether – the class
Reptilia, in fact. Once again, she leads me to a place where
few would even pause to look. Accessed via a road that
leads to the Chelsea Women's Football Club (rather more
modest in appearance than the men's) and behind a car
park is an area of unremarkable scrubland, around half
a mile from Kingston town centre. There are AstroTurfed
pitches to one side and blocks of flats and a factory to the
other. A broad, sturdy-looking turkey oak stands towards
the back of the scrubland, its leaves beginning to turn
from lime-green to yellow and orange. We can hear the
sound of birds singing and bees buzzing, and I can hear
crickets ('Can you really?' quizzes Alison, explaining that
she lost the ability to hear at this frequency some years
ago).

Alison stops to show me a flowering plant called
buck's-horn plantain with slender, poker-shaped spikes,
its leaves the shape of a stag's antlers. 'It's one of my
favourite pavement plants,' she tells me. 'You don't find
it everywhere – it likes poorly drained areas – but you see
it's pollinating.' She has previously recorded common blue
butterflies here too, she adds.

Alison has been using a car park next to this piece of
land when working on a site along the river as part of
her day job. It first caught her attention when she saw

a post by one of the local councillors on social media, announcing that the council had found a patch of scrubland which they could use to build nineteen new houses; 'And I thought, oh really?' Alison relays with an air of mischief; 'Better go and have a look!'

The first thing she noticed was that she wasn't the first person to check out the land – someone had already put down the felt mats used in ecological surveying to detect the presence of reptiles, much like those I encountered in my search for great crested newts. So, naturally, when Alison saw the mats she decided to peer underneath, 'and sure enough, I found a number of slow-worms!' she tells me triumphantly.

Slow-worms are not worms at all (nor are they particularly slow, once they get going). Often confused for snakes, they are in fact legless lizards and can be differentiated as such in a number of ways. Firstly, at 40–50 centimetres in length they tend to be smaller than snakes. Secondly, unlike snakes they have no neck, with their heads running seamlessly into their bodies. Thirdly, they have eyelids with which they can blink. Fourthly, their tongues are notched rather than forked, although admittedly this one is harder to see. And finally, although they have scales they look visibly smoother than most snakes (with the exception of the aptly named smooth snake). This is down to the fact that their scales do not overlap, a feature that probably evolved from their fondness for burrowing into the loose soil and decaying vegetation beneath logs and rocks. Like snakes, slow-worms shed their skin as they grow, but they slough it off in patches rather than all in one go, so if you find smaller flakes of reptile skin they most likely used to belong to a slow-worm.

Their tendency to burrow is also probably why slow-worms lost their limbs. Although they might be the only species of legless lizard in Britain, across the globe there are plenty of lizards that have evolved to have only tiny legs or no legs at all.

Whoever had previously been surveying here wasn't up to Alison's standards – she describes the work disparagingly as a 'quasi-ecological survey'; by July the mats were gone, too soon in the year to determine whether a breeding population exists in the area. So she decided to put down some mats of her own. She warns that we may not find any slow-worms today, as it's close to hibernation time. 'Tomorrow I'm taking all my mats away,' she adds, 'and that's the end of my little survey.' In October, which begins in just a few days' time, the slow-worms will tunnel underground to hibernate and they won't come up again until spring.

Alison tells me to lift my feet carefully as I walk, so I don't create too much vibration around the mats, which could end up spooking the slow-worms. She has placed the mats strategically around the edge of the land, close to some dense patches of bramble. Gingerly lifting the first mat, she declares it unoccupied, but as Alison has previously advised me, her eyesight isn't as good as it once was. In fact, amongst the leaf litter is a large female, her smooth, golden body curled into a back-to-front S, gleaming in the autumn sunshine. Her sides are a rich dark brown, her geometrically patterned scales lying perfectly flat, with a thin line running all the way from the back of her head down to her tail. The males lack this line down the middle, as well as the darker sides. They are smaller and often greyer in colour, some with a smattering of bright-blue specks across their backs.

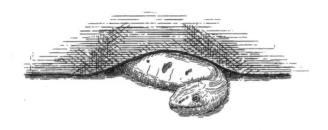

Once she realises she's exposed, the slow-worm quickly burrows downwards, although burrows isn't quite the right word – it looks almost like she's swimming. It's no wonder she makes a hasty exit – slow-worms are vulnerable to predators, of which they have many, including adders, hedgehogs, badgers, various bird species and domestic cats, which will merrily destroy every slow-worm they encounter. Slow-worms are, however, able to 'smell' the presence of predatory versus harmless snakes by flicking their tongues in and out, and they have some unusual defence mechanisms. If fleeing isn't an option they will try defecation, producing a scent unpleasant enough to deter some predators. That failing, they have another line of defence known as 'autonomy'; like other lizards (but not snakes) they are able to detach their tails, which continue thrashing about for several minutes, distracting the predator while they make their escape.

The scientific name for slow-worms, *Anguis fragilis*, refers to this ability, *fragilis* meaning fragile. Some studies have found that between 50 and 70 per cent of wild slow-worms have lost their tails. Once the tail is gone the slow-worm can look a bit stumpy, but over time it regrows. This is, however, a costly biological process and the new tail is usually slightly shorter than before, never quite returning to its former glory. It's also a tactic that can only be deployed once – the new tail is no longer detachable.

Alison and I replace the mat carefully, so as not to disrupt the slow-worm any further, and move on to the second mat. After our early success, this time we find nothing. Alison is a bit disappointed – she was hoping we'd see some of this year's young. Slow-worms mate in May, with a rather unorthodox courtship. The amorous male bites on to the back of his chosen female and the pair entwine their bodies, rolling and writhing for up to ten hours, sometimes leaving the female with scars (and presumably leaving both exhausted). Unlike most reptiles the female gives birth to live young, around eight in total, which hatch while they're still inside her body, living off the yolk of the egg. They emerge in late summer, about 4 centimetres in length, still wrapped in their embryonic membranes.

There are only two mats left, Alison tells me; one must have got nicked. The third is as empty as the second. 'It's a shame,' Alison says, 'because that one slow-worm is likely to be the last one you see this year.' I bend down and cautiously turn over the fourth and final mat. Underneath are not one but two much smaller slow-worms, silver in colour, with black stripes down their middle, a black dot on their heads and sleek dark sides. They're more wriggly and less refined in their movements than the adult female. Alison takes a quick photo on her phone before replacing the mat.

As we're about to leave, I spot the fifth mat, which hasn't been taken after all. Lifting it up, we count a total of six young slow-worms resting underneath. Unlike the previous two they take time to sense our presence, slowly starting to move as they feel the heat of the sun on their backs. One by one they disappear through the grass and into the ground beneath, as if sucked downwards by the earth itself.

'Unbelievable!' Alison exclaims. 'That's the best result so far for hatchlings, under one mat. You are totally and utterly lucky!' She puts our success down to the unusually sunny weather, and it's proof, she adds, of a breeding population. 'We go to some really nice places and put down loads of mats,' she observes, 'thirty or more in an area of this size, in suburban areas too, not urban like this one, and we don't get anything. And we wouldn't have a clue they were here unless we'd done this study, would we?'

It's very true – in fact, plenty of people don't know

that slow-worms even exist as a species, let alone on their urban doorsteps. As creatures that spend most of their lives undercover they're easy to miss unless, like Alison, you're deliberately looking (or you're going through a compost heap, where they are often found). 'I don't walk past anything without investigating,' Alison confesses; 'even a road cone, I'll pick it up and find a wood mouse underneath.'

This general inquisitiveness has made Alison a few enemies as well as friends, especially on the local council. 'They aren't too happy with me, because they keep granting themselves planning permission and then I point out everything they've done wrong,' she says with a wry smile. She's eagerly awaiting the planning application for this particular site, which she will scour to make sure slow-worm mitigation has been included. If not, she now has enough evidence to prove there's a breeding population here, although it's the number of adults that dictates the conservation status, she explains, and whether this counts as a low-, medium- or high-density population. 'I'd say this is a medium population,' she argues, 'if you consider the numbers that we're getting elsewhere.'

'In the meantime,' she adds, 'I've written a naughty poem, which gives them a clue. "Will the slow-worm turn and protect biodiversity", that's the last line.'

Alison's dedication is remarkable; I wonder how often a nondescript area like this gets built over, no one ever knowing what was here in the first place, and then it's gone forever. Alison nods, pointing at the block of flats which overlook the scrubland. 'They're new too, everything's changing so quickly,' she sighs. Slow-worms are thought to be one of the longest-lived lizards on the planet. Although hard to substantiate, their average lifespan is believed to be between twenty and thirty years, and in captivity one slow-worm in Copenhagen Zoo reportedly lived to over fifty-four years old. It's certainly long enough for change to take place many times over in an urban area like this, where councils come and go and building programmes continue apace. Hopefully these slow-worms will at least survive through some of that change with Alison lobbying on their behalf.

DIGGING DEEPER

On a brilliantly sunny April weekend the city of Brighton and Hove buzzes with life. There's a festival atmosphere in the air; people sit outside cafés and bars spilling on to the beach, while buskers strum their guitars and screams emanate from the pier where families ride the rollercoaster.

Meanwhile, a metre or so beneath the surface of the city, in chambers connected by networks of tunnels, sleep Brighton's many badgers, waiting for the cover of darkness before emerging in search of their dinner.

Tim Roper is one of the few people in the UK to have studied urban badgers. A couple of years into his lectureship at Sussex University, while walking in the South Downs – the chalk cliffs that run along the south-east coastline – he began to find evidence of badgers. At the time, most existing badger research had taken place in Scotland, but the Scottish landscape, with its vast stretches of woodland and countless hiding places, made it hard to observe their behaviour. The downlands weren't considered to be typical badger territory, but, knowing they were here, Tim supposed it would be possible to track and watch them far more closely. And as it turned out, he was right.

Tim soon learnt that the badgers were venturing beyond the South Downs and much further into the city itself. Or, in fact, it was a combination of some badgers coming in and others staying put as Brighton was built around them.

Tim and I meet at the top level of the Brighton Marina car park, which looks straight out over the sea. We walk the short distance to Sussex Square, a private, gated gardens surrounded by Grade I listed townhouses with roof terraces and sea views. Of course, we're not expecting to see any badgers at this time of day – what we're looking out for is their setts, one of which is located in the centre of the gardens. Unfortunately for us, the well-heeled local residents refused to lend Tim a key, so we peer through the railings instead. We can't quite make out the sett itself, but in the middle of the freshly mown lawn I catch the eye of a young fox enjoying exclusive access to the gardens, its fur gleaming in the sunlight.

'This is some of the most expensive housing in Brighton,' Tim tells me, 'but four minutes' walk and you'll find the cheapest.' The badgers, he explains, appear equally happy both here and in the nearby council estate.

We continue beyond the square, along a street of slightly more modest terraces, until we reach a dead end, with rows of garages and a stone wall behind. Over that wall, Tim says, is another sett. Or at least there was, when he was last surveying the area – ten years have now passed and he has since retired. But a decade isn't long in the lifespan of a badger sett. Setts are often passed down through generations of badgers, some reaching hundreds of years old. 'If you look on a map and see something with badger in its name,' Tim remarks, '"Badger Hill", for example, then quite often you'll find a badger sett in that spot which presumably has always been there.'

Setts are seldom much deeper than a metre or so underground (in this country at least – in colder climes badgers dig deeper). But they can be huge in terms of the space they occupy – hundreds of square metres or more. The larger the sett, the older it's likely to be. Tim is one of the few people to have excavated badger setts, during the building of the Brighton bypass. One of the setts he explored had over 100 entrances and numerous different rooms connected by metres of tunnel.

I try to imagine what it must be like to spend so much time below ground, in the bowels of the earth, living off a diet of worms. I suggest to Tim that if I were to be reincarnated as another urban creature, I would far sooner be a peregrine than a badger. But Tim advises I might be surprised by the badgers' living quarters, which he found to be both pleasant and commodious. The setts were clean and sweet-smelling, he tells me, with a separate en suite for the cubs. The nest chambers were padded with hay and straw and equipped with a damp-proof lining of empty fertiliser bags that the badgers collected from the fields. 'It really was a bit like *The Wind in the Willows*,' he jokes.

Badgers are master craftsmen, their setts an incredible feat of engineering designed with mechanical efficiency, which is all the more baffling when you consider that they work in absolute darkness. Somehow they manage to get two or even three tunnels to meet, with the angles exactly right. We still have little idea how they accomplish this, Tim concedes.

In urban environments setts don't last as long, because there's much more disruption around them. But since this particular sett lies in a piece of no-man's-land, Tim believes it could have been here for a long time. Surrounded by brambles, small trees and dense thicket, it's typical of the kind of place where you find urban badger setts.

It was the living arrangements in setts like this one that sparked Tim's interest in badgers, he tells me. It wasn't until the 1940s, when a schoolteacher named Ernest Neal began sitting outside badger setts, that it was discovered that badgers live in groups. 'If you look at literature up until the middle of the last century,' Tim notes, 'in children's stories and so on, badgers are always represented as solitary, grumpy old creatures.' Although they live communally, they usually sleep in separate chambers, except in the winter when they huddle up for warmth. And, unlike other cohabiting animals, they don't seem to be particularly social. When they go off foraging they do so alone, unless it's a mother with cubs, and they aren't known to show cooperative behaviour. 'They seem to be a sort of relatively early step on the way to socialisation,' Tim tells me.

The largest number of badgers to a single sett was recorded by scientists in Wytham Woods in Oxford – between twenty and thirty individuals – but the Wytham Woods badgers are a special case, both fed and protected by those that study them. More typically a sett is likely to house what Tim describes as 'the basic unit' – a male, female and their cubs, or young adult offspring, until they reach around eighteen months of age.

The cubs will leave home as soon as they can to start out on their own or take over an existing sett. But, like the human property market, land is at a premium, forcing many young adults to stay at home with mum and dad until a vacancy opens up. 'What they seem to be doing is playing a kind of waiting game,' Tim explains, 'so if an old animal from a group next door dies, then a younger animal from a neighbouring group will try and come in and take over that status.' This can result in fearsome battles, as young males and females fight for breeding status. The winning couple become the dominant animals, and in most cases only they are able to become parents within the group, so the ability for badgers to pass on their genes is far from guaranteed.

In common with other members of the mustelid family, which in the UK includes stoats, weasels and martens, badgers have rather unusual breeding habits. They'll give birth as early as February and often mate immediately afterwards. In a process known as delayed implantation, the fertilised egg will sit in the female's uterus until November time, before it finally implants. This process has probably developed through evolutionary and selective pressure, Tim explains: the young need to be born as early in the year as possible, giving them the maximum amount of time to gain weight before winter. In colder climates, badgers hibernate fully, whereas in Britain they remain active, but they don't wander far from the sett and their food intake drops drastically.

We move on in search of the next sett, heading towards a busy main road and passing a large construction site next to the Royal Sussex Hospital, where the skeletons of three new tower blocks loom. When Tim was surveying this area, the works had yet to begin and there was a badger sett under one of the Portakabins. The security guards admitted to feeding the badgers and a local nurse informed Tim that a badger once followed her down a hospital corridor on one of her night shifts.

While this is far from typical badger behaviour, it does seem that urban badgers are adapting to life in the city. In rural settings they are highly territorial, marking out their patch with latrines. But in urban areas latrines are rarely found, and Tim suspects they don't feel the same territorial pressures. They must also have grown habituated to the sensations of city living: the vibration of traffic above their setts, the constant light pollution and the pungent scent of humans.

Like most successful urban exploiters, such as foxes, magpies and squirrels, badgers are generalists and will eat pretty much anything. They wander around with their noses to the ground and gobble up whatever they encounter, be it plant or animal. The bulk of a rural badger's diet is made up of earthworms – up to a kilo on a good night, when the ground is damp and the worms are drawn up to the surface to feed on dead leaves and vegetation. Slurping them like spaghetti, the badgers inflict one or two fatal bites before swallowing the worms down whole. They'll also eat insects, seeds and berries, eggs, carrion and smaller, slow-moving creatures like toads. (A study in France found that during large-scale toad migrations, the stomachs of local badgers were almost entirely filled with the unfortunate amphibians.) Rural badgers are, sadly, voracious predators of hedgehogs too, which they flip upside-down to avoid the spines and ruthlessly rip open.

Urban badgers will hoover up all of the above, with an extra portion of fries; but instead of heading straight out at dusk like their countryside-dwelling counterparts, they'll wait until later to avoid crossing paths with too many humans. Despite this reduction in foraging time, they weigh on average 1 kilo more than their rural relatives, which just goes to show the culinary benefits of city living. They are probably drawn towards human food by the sugar and fat content (which is, of course, what makes it so delicious). Far from watching their waistlines, badgers need to gain inches of fat beneath their long, shaggy coats ahead of each winter. 'I suspect once they discover a new kind of food they pass this information on across generations,' Tim asserts; 'the young follow their mother for most of the first year, and they'll trundle around eating whatever she's eating.' I ask whether, given the choice, they'd pick worms or pizza. 'Probably the pizza!' he laughs.

Many people deliberately feed them too; a friend tells me her grandparents leave out peanut-butter sandwiches, and while studying the area Tim came across one particular resident who would cook two dinners every night – one for him and one for the badgers. His speciality was pot noodle, which isn't too far removed from worms, I suppose.

'One of the things that interested me when we started working on urban badgers,' Tim recalls, 'is the fact that the conflicts are so extreme. You get one person that's feeding the badgers in their back garden, while their next-door neighbour is getting their garden dug up and is absolutely furious about it.' For this reason Tim advises against leaving out food. 'Although they're naturally wary of humans, I've seen photographs of badgers eating in people's kitchens. But no matter how tame they might seem I would never trust one myself – they have very sharp teeth and powerful jaws and once they grab hold they won't let go. I would only handle one once it had been anaesthetised!'

As our sett tour continues, we pass the prestigious Brighton College boarding and day school, which houses another sett in its sports field. We turn right, up a road with a steeper incline, with houses on our left and a wooden fence on our right. Behind the fence is a slice of land and a steep drop, where a chunk of the Downs has been scooped out to make way for an industrial estate. Another stretch of no-man's-land that the badgers have commandeered, the bank is lined with small trees, a mass of unruly ivy and a few clumps of early bluebells. Tim hunts for a gate that no longer exists, but we do spot a badger-sized gap at the bottom of the fence which looks suspiciously like it has been made deliberately.

While we're unable to see the sett from our limited vantage point, a number of pathways are clearly visible, formed as the badgers trundle up and down on their way to the Brighton racecourse, which provides an excellent worming ground. As both subterranean and nocturnal creatures, badgers have small eyes and poor eyesight (hence why they're often pictured wearing spectacles in whimsical illustrations). They therefore depend on their noses to navigate, and use the same routes when travelling to and from the sett, like trains on a network of railway tracks. Once they've reached a suitable foraging ground they'll wander off the track, eating their fill before hopping back on until their next stop.

We reach a small park where Tim shows me the entrance to yet another sett, partially hidden in a bramble patch. It's narrow in circumference – designed to snugly fit an adult badger. The earth around the hole has been compacted, smoothed and polished through regular comings and goings.

Passing on through the park, Tim wants to show me one more sett before we leave. Having carefully mapped our route, and armed with a printout (he's got a smartphone, he says, but hasn't yet learnt how to use it), we spend a short while looking for a path between the rows of terraced houses. Ten years having passed since he last visited, we start to wonder whether the land has been developed and the sett no longer exists. Just as we're about to give up, Tim spots a section of dirt track behind some gardens. We follow what looks like a makeshift path down a steep bank. This bank is ideal for sett-building, Tim explains; the soil is well drained and the slope allows the badgers to get deeper while digging horizontally – much lighter work than digging diagonally into a flat surface.

One of the locals seems to have removed a few panels in their back fence and is using the bank to grow vegetables – a sign, Tim fears, that the badgers are no longer present, or they would have quickly demolished the vegetable patch. As we're still deliberating, another resident pokes out his head from above a fence, warning that we can't pass along any further. Tim explains that we're looking for an old badger sett, to which the resident replies that he's seen badgers here on occasion, pointing to a nearby entrance that we somehow managed to miss. The hole is quite overgrown, so unlikely to still be in use; Tim guesses that the badgers have added an extension and moved further along. Where badgers move on, foxes will sometimes move in, he adds. Larger setts can even accommodate badgers at one end and a family of foxes at the other, which the badgers will just about tolerate.

As we head back to the car park, Tim reveals that in the area where we've been walking his team discovered eight to ten setts, the highest density ever recorded per square metre. Although this feels like the heart of the city it's still connected to the downlands, giving the badgers plenty of options for dinner: worms on one side, fast-food chains and gardens with friendly locals on the other. 'Over time, some urban species are becoming commensal,' Tim observes, referring to animals like rats and mice that are reliant on humans for food and shelter. 'These days you could describe foxes as such – they too have become dependent on humans. Badgers haven't yet reached that stage, but they're on the way. It's now largely a matter of how well we humans will tolerate them.'

Later that evening, Ben drives the two of us to Seaford, a coastal town 13 miles east of Brighton. Sunset isn't until eight o'clock, but we've arranged to arrive a bit earlier so that we have time to get into position for the evening's activities. Although it's only about 8 or 9 degrees there's very little wind in the air and we're warmed by the last glow of a day of uninterrupted sunshine.

We park outside the house of Bryony Tolhurst, a Senior Lecturer in Behavioural Ecology at the University of Brighton. Before we head off in search of more badgers, Bryony wants to show us a sett less than a five-minute walk away. It's located on a small scrap of wasteland that slopes steeply upwards surrounded by houses and flats, as nondescript a piece of land as you can imagine. Hardened patches of ground reveal the earth beneath, surrounded by scrub, clumps of grass and knots of brambles and ivy. The land is owned by a developer and Bryony has been told by her neighbour – a local Green Party councillor – that an eye-watering £1 million was paid for it, with plans to build a new block of flats. 'People have this idea about pristine wildlife habitats,' she observes, 'but actually some of the best places for wildlife are these sites where most people would never even think to look.'

Bryony had her suspicions that there was a sett nearby after spotting signs of badger activity, including well-trodden paths on a neighbouring bank. 'Along the top of that bank I caught a whiff of badger,' she tells us. 'I don't even have a strong sense of smell, but since badgers use scent communication they absolutely stink!' She has learnt to distinguish between the aromas of different mammals – of badger, fox and dog. 'When I used to go out in Brighton, I'd play a game of "fox or ganja", she jokes, 'if you get a strong smell on the wind, it's either going to be a fox or someone smoking a joint!'

Bryony's detective work led her to this site, where she saw signs of excavation that had nothing to do with the developers. The badgers, it transpires, decided to move in shortly after the land was purchased. This puts the developers in a tricky position – it's illegal to build within 30 metres of a sett, so their only option will be to work with Natural England to move the badgers at considerable expense, or give up on their plans to build altogether for the time being. 'I mean, they might try and get away with bulldozing the sett,' Bryony adds, 'but that would be breaking the law.' I point out that with a mammalian expert living just around the corner they are also unlikely to get away with it. 'Yes, I'm the badger guardian!' she jokes.

Bryony waits for a few passers-by to leave before we scramble up to take a closer look at the sett, as this is technically trespassing. The first hole is clearly active. Outside it sits a pile of dried grass, which is probably bedding material that the badgers are airing (for such smelly creatures they are very fastidious). We scramble across to another entryway.

Piles of rocky white Sussex flint lie at the mouth of the hole where the badgers have been digging. 'In the countryside I'd be really careful and not go right up to a sett like this,' Bryony explains, 'but I imagine urban badgers can't be that bothered by human scent – I know urban foxes aren't.'

With sunset fast approaching we head back to Bryony's house, briefly stopping to take a peek at her small garden, which she has deliberately left to grow wild and equipped with swift boxes and a small pond. In the past she has put out camera traps, so she knows that the badgers sometimes visit. We jump into our respective cars, Ben and I tailing Bryony as we drive towards the edge of Seaford. We pass a golf course and turn up a road that cuts through a couple of fields, leading to a nature reserve ringed by coastline. Strictly speaking, this location is suburban at best, but having surveilled the sett near her house Bryony soon learnt that the badgers living there don't venture out until well into the early hours. So instead she has chosen a sett where she hopes we might catch a glimpse of the occupants as they begin to emerge at dusk.

Sett-watching is something that anyone can do. It simply involves sitting very still, downwind of a sett as it grows dark, in the hope of spotting the badgers during the short window when it's dark enough for them to come out, but there's enough light left for us to see them. You don't need any fancy equipment, just a torch to find your way back.

We stroll towards the top of a hill, with houses behind us and sea in front. During the day, Bryony tells us, this area is rammed full of cars transporting visitors to the nature reserve, but now it's incredibly peaceful, the only other vehicle a white van parked a bit further up. As the sun descends below the waves it casts a dazzling glow of yellow, orange and pinky-red, its rays extending like fingers over a field full of sheep. At the crest of the hill we watch the silhouettes of half a dozen rabbits as they gambol about; it's almost too perfect a picture – Ben and Bryony joke that we've found ourselves in *Watership Down*.

Bryony checks that we're 'OK with a bit of mild trespassing' before directing us to climb over a rusty farm gate. Together, we tread carefully through an area of dense undergrowth. She uses her finger to test the wind direction, making sure we're downwind of the sett so that our scent doesn't startle the badgers. We settle on a spot beneath the branches of a few small trees and bushes, next to a barbed-wire fence, behind which we can see two openings to the sett, marked by the little piles of Sussex flint where they've been excavated. We make ourselves as comfortable as possible – I choose to sit cross-legged on a patch of soft green moss. This initially seems like a good decision until I notice the stinging nettles, which somehow manage to sting me through two pairs of leggings.

I keep my eyes fixed on the two sett entrances, determined not to move, while Bryony scans the clearing behind us. As I begin to tune in to our surroundings I realise this is the most peaceful I've felt in weeks. A blackbird starts singing and my nostrils fill with the scent of fresh vegetation. Having dragged Ben along, I was worried he might find my proposed evening activity boring, but he seems perfectly contented.

I listen out for any signs of movement, ears primed for the rustle of a leaf or the snap of a twig. Suddenly and silently a fox appears, its fur a deep, luxurious red. We must be well camouflaged, because as my eyes lock with the fox's we seem to have caught each other by surprise. For a moment I gaze into its deep-hazel eyes, before it disappears as swiftly as it arrived.

As the light starts to fade, the proportions of the shapes around us warp and transform as my eyes adjust. Shadows appear and a sliver of moon emerges above our heads in the dark blue sky.

We wait, and wait, not wanting to admit defeat until Bryony breaks the silence, conceding that it's now too dark to see much anyway. A short while ago, she says, she caught that unmistakable whiff of badger at the same time as something startled the birds – but then it was gone.

Bryony fears that the group occupying this sett may have been seriously diminished after a couple of dead badgers were recently spotted on the road nearby. Death on the roads is one of the biggest threats that badgers face; in fact, the only way most people encounter a badger in real life is when it becomes roadkill. Badgers still suffer from human persecution too – both through badger-baiting and at the hands of farmers that shoot them without a licence. Populations have also been affected by legalised badger-culling, a practice that is being slowly phased out by the government that first introduced it, after – as many scientists predicted – it has proven to have little impact on case numbers of bovine tuberculosis. Of course, shooting and culling are mainly 'countryside' issues which urban badgers are less likely to face.

Bryony takes out her torch and suggests we head back to the road and over the top of the hill to see if we can find any badgers out foraging instead. We carefully watch our step – both to avoid tripping and so as not to startle the badgers. Briefly pausing, we look back at the houses lining the bay below and the ships out at sea; Bryony points to the bright lights of a giant ferry. Over the hill we follow a path that leads towards another sett. My senses are heightened by the darkness – I can smell the coconutty scent of the gorse bushes and taste the salt in the air. Bryony sweeps the torch from side to side, focusing on the path and the clearings – it's not necessarily where the badgers are most likely to be, she explains, but it's where we'll be able to spot them. We could do with a more powerful flashlight, she adds, but that would be more invasive and something for which you should really seek permission.

What Bryony is looking out for is eye shine, common to most carnivores, though not humans; when a pair of eyes glint through the darkness, reflecting back the light of the torch. (If you've ever taken a photo of a pet dog or cat with a flash, you'll likely have captured this eerie, slightly demonic look.) The eyes of different species shine with a slightly different colour and brightness, allowing her to distinguish between foxes, of which we soon spot a couple, and badgers.

As we get closer to the open sea and the sound of the waves grows louder, we reach a clearing. I'm just behind her as Bryony calls out 'Badger!' excitedly. She identified the beast by the glow of its eyes, but I just catch its behind – a grey blur, quickly disappearing into the other monochrome shapes that surround us.

We sit down in the clearing for a moment, to avoid spooking the badgers any further, before heading back along a different path that leads us through an area of woodland. Every now and again we startle a sleeping wood pigeon, jumping out of our skins at the loud whoosh and heavy flapping of wings. As Bryony continues to sweep the torch, the eyes of another badger briefly flash. Or perhaps it's the same badger as before – it's impossible to tell. 'Well, at least you saw a badger's bum!' Bryony laughs as we head back to our cars.

IN THE SHADOWS

Some creatures are compelled to burrow and dig into the soil itself, a strong biological force pulling them downwards into the warmth and relative safety of the earth. Others are cave dwellers, seeking out the darkness and stable conditions of voids within the ground. Today I'm in search of the latter.

The landscape of Birmingham and the Black Country has been shaped beyond recognition by its industrial past. Huge lime kilns, vast foundries and coal mines, remnants of former industry fallen into disuse, become part of new developments, or are reclaimed by nature itself. As we learnt in Chapter 4, the lifeline that connected all of this industry was the canal network, which was used for transporting the heavy limestone, coal and sandstone to the places where it was processed.

'It's said that Birmingham has more canals than Venice, which is true if you're counting the number of miles,' Morgan Hughes of the Birmingham and Black Country Bat Group informs me as we walk along the towpath. 'My great grandad lost both of his legs in one of the mines here,' she continues, 'and my grandad, his son, was head of one of the foundries. So I'm a real local.' Scott Brown, another member of the group who has joined us this evening, adds that he comes from a family of engineers, otherwise known as canal navvies. 'I think that a lot of the people here have strong ties to their industrial past,' Morgan asserts, 'and now, some of our best sites for bats in the county are related to former industry.'

It's a cool, late-September evening and Morgan, Scott and I are headed to the Netherton Tunnel, a short walk along the canal from Dudley Station. The canal itself is thick with rubbish: takeaway boxes, beer cans, plastic bottles, a child's football. Dudley is an area of extreme deprivation with half its children living below the breadline, Scott tells me, so understandably litter-picking isn't a top priority.

We soon reach the entrance to the tunnel, a perfect half-circle of grey brick that joins up to form a whole with its reflection in the water. Grass, brambles and scrub cover the length of the banks and spill over the top, with towpaths running along both sides. Built to connect the towns of Netherton and Tipton, the tunnel was first opened in 1858 and was the last of its kind to be built in Britain during the 'Canal Age' of the nineteenth century. It was designed to relieve the bottleneck of the adjacent Dudley Tunnel, which could only fit one boat at a time. Back then, those piloting the boats had to lie on their backs and push through the tunnel using their feet, a slow and arduous process that resulted in long queues, taking several days to get through in some instances.

The Netherton Tunnel was made wide enough to accommodate two boats, with towpaths running either side so that horse-drawn narrow boats could be pulled through. The tunnel, 2,768 metres in length, was initially lit using gas, then later by electricity, and now it's no longer lit at all, 'which is great for the bats!' Morgan smiles. (Astonishingly, this doesn't appear to deter a cyclist without any lights on his bike from using it as a shortcut. He pops out just as we reach the entrance to the tunnel, waving cheerily as he rides past.)

Tonight we're looking for Daubenton's bats, otherwise know as 'Daubs', 'bennies' or 'water bats', which have a strong affinity to water. Up close they have pinkish faces, bare-skinned around their beady black eyes, with fuzzy brown fur, silver-grey bellies and large, five-clawed feet. Outside of the hibernation period, they use the tunnel as a roosting site and a corridor for commuting. Its proximity to good food sources makes the tunnel particularly appealing; Daubenton's bats are trawlers – they catch insect prey on the surface of the water, either directly into their mouths or by scooping them up with their tails. The tunnel is flanked by two nearby nature reserves – Bumble Hole and Sheepwash – both of which have plenty of pools, filled with insects. Morgan and Scott suspect that this is where the bats are headed as they venture out each evening.

Morgan deliberately suggested we meet ahead of it getting dark so that we can take a look inside the tunnel before the Daubenton's begin to emerge. It's 6.30 p.m., and since the clocks have yet to change we've got half an hour before sunset. Inside the pitch-black tunnel the air smells of damp; there's a constant drip, drip, drip from the ceiling above and a squelching beneath our feet as our wellies crunch on the gravel towpath. Morgan tells me that when she last walked the length from one end to the other it took her close to an hour. A bat, on the other hand, can fly through in ten minutes.

Both Morgan and Scott are professional ecologists by day – Scott for a consultancy and Morgan for Network Rail – but they spend their evenings out looking for bats with groups of volunteers. Between April and October, Morgan is out five nights a week and Scott two, surveying on weekdays and trapping into the early hours at week-ends. 'This is our life,' Morgan tells me, 'and we've all been friends for years, so it's our social life too. I don't sleep a lot anyway.'

During the winter, they conduct hibernation surveys inside the nearby man-made caverns, donning wetsuits in the snow and paddling in inflatable kayaks as they search amongst the cracks and crevices for torpid bats. In a matter of days or weeks, as the weather starts to turn the bats will begin to settle in, their heart rates dropping from up to 1,000 beats a minute while in flight to just two. Even when it's freezing outside, it is humid inside the caverns, the temperature is fairly stable and it's generally free from predators – the exact conditions bats need for hibernation. Morgan also leads endoscoping surveys: 'It's a little camera with an LED light on the end of a flexible tube, which you insert into the cracks to find hibernating bats,' she describes; 'it's like what doctors put up people's bums, but without the petroleum jelly!'

Until recently, it was believed that Birmingham and the Black Country had little to offer in the way of bats. In 2010, Morgan explains, the Wildlife Trusts produced a bat rarity index that listed twelve species in total, of which eleven were thought to be present in the county, but only pipistrelles and noctules were believed to be common. Morgan and Scott instinctively felt that this was wrong and that the issue was a lack of data. They've since focused their survey work on areas like the Netherton Tunnel to counter the assumption that biodiversity is low in urban areas. 'We've been on a bit of a mission since then to improve the quality and number of biological records,' Morgan explains, 'training dozens of volunteers over the years. The main thing is trying to debunk people's misconceptions about bat diversity in urban areas.'

Bats are acutely sensitive to light and noise, so they can face many obstacles. Roads and motorways in particular create significant barriers, as do electric lights, the results of which can be extreme – if an area becomes too bright, some species won't come out from their roosts at all and can end up starving to death or becoming entombed. But the Dudley Daubenton's must have grown habituated to a certain level of disturbance, with the benefits of the tunnel and access to waterbodies outweighing the downsides.

Morgan and Scott have also been gathering evidence that the canal, along with other remnants of former industry, is helping bats overcome those barriers. Through their surveying work they have changed the status of most of the twelve species featured in the bat rarity index from rare to frequent, or frequent to abundant. 'Ecologists often have to spend their time looking at crappy buildings for their day job, so when they're volunteering, they want to go out to more rural places to find the "sexy bats", Scott jokes, 'but the species we need to conserve are often found in urban areas.'

Of course, most people don't find any species of bat particularly sexy, which is why Morgan and Scott devote so much time to conserving them. 'They're in desperate need of conservation,' Morgan explains, 'and they're vilified by the media, cinema, literature – there's a lot of fear around them . . .'

'Bram Stoker's *Dracula* was to bats what *Jaws* was to sharks,' Scott interjects.

This hasn't been helped by reports that the coronavirus pandemic began with Chinese horseshoe bats, and the resulting belief that bats are vectors for viruses and disease. It's true that they have one of the most robust immune systems of any living animal, meaning they can carry viruses with no noticeable impact on their own health (they're less prone to developing cancers too). But they are unlikely to pass anything directly on to humans – we're not closely enough related. Scientists believe that COVID-19 wasn't given to us by bats directly, but jumped across from another intermediary species – the virus was detected in both pangolins and civets, creatures that are often found in the wet markets where COVID-19 is thought to have originated. Ironically, during the first lockdown, specialists like Morgan and Scott were advised to avoid handling bats in case humans could pass COVID-19 back to them, with unknown consequences for our British bats.

Having walked a few hundred metres into the tunnel, Morgan finally finds what she's looking for – a chink of light in the roof. 'There,' she points, 'it's one of the pepper pots!' Named after their shape, the pepper pots are brick-lined openings that run along the length of the tunnel at 350-metre intervals, providing ventilation. This allows modern boats that pass through the tunnel – these days mainly for leisure – to use their onboard engines, something that is prohibited in many canal tunnels.

Morgan is curious as to whether the bats use these air vents to travel in and out of the tunnel. She and a friend once walked all the way through, she tells me, and then back over the top by following the line of pepper pots. 'There's one in someone's garden and she was just coming out of her house as we passed, so we asked her, "Do you see bats coming in and out?" She said no, she'd never seen anything, but the thing is, people don't look.'

Most of the time when she's out surveying people pay little attention, Morgan claims; 'Although a few weeks ago we were putting up nets and someone asked, "You're not catching ghosts are you?"' At the end of a talk by the Birmingham and Black Country Bat Group, she adds, a member of the audience came up and said, 'Thank you so much, I appreciated that, I didn't realise that bats were real.'

Morgan suggests we start heading back, as the bats will soon begin to emerge. On the way, Scott stops to show me the kind of spot where they might be roosting – a tiny gap in the mortar of a cavity wall. At 7–12 grams in weight, Daubenton's bats are a fair bit larger than the tiny 5-gram pipistrelles I encountered in Chapter 2, but as crevice dwellers they can flatten themselves to the point where they only require the width of a pencil to squeeze their way in.

As we come out of the mouth of the tunnel Morgan

goes to find Tamar, an outdoor education instructor and regular volunteer. Together they've spent the past few months conducting emergence surveys, with someone stationed at each end of the tunnel, counting the bats that go in at one end and come out of the other over a one-hour period. 'During emergence time the bats that come out minus the bats that go in gives us the number that are resident within the tunnel,' Morgan explains. By her reckoning, this current roost is over seventy-strong.

Morgan gets out a red LED torch, explaining that red light bothers the bats considerably less than white. She places it so that its beam cuts across the canal, illuminating the bats as they fly through the light. 'It's the same with all mammals,' she adds; 'if I shine a red light in your eyes, you won't find it as harsh as a white one.'

If you want to go out and see bats for yourself, most local bat groups lead regular walks, but it's possible to see them without any fancy equipment too, Morgan tells me. She suggests an evening trip to your local waterbody; 'You can go to pools and watch Daubenton's foraging, which is always spectacular,' she enthuses, although it's important to exercise caution to avoid disturbing the bats. Only those with the proper licence are allowed to search for roosts and technically only professionals can use a torch like we have this evening. 'If your intention is to look for bats and you're knowingly causing disturbance then you need a licence,' Morgan explains, 'but if you were a fisherman sitting here with your LED torch, it would be OK. So it comes down to the interpretation of the law. I don't think anyone would throw the book at you for sitting with a red torch for a couple of hours.'

The team tonight have bat detectors, which will confirm the species for us, but the way they fly is also a big giveaway. While pipistrelles flit and dart high along the tree lines, the Daubenton's flight is direct and purposeful. They fly low, hugging the surface of the water like tiny hovercraft. This allows them to skim off small insects, like midges, caddisflies and mayflies, while also protecting them from being predated themselves by owls or domestic cats. ('They're small and full of muscles, so they're very tasty,' claims Scott, adding that he's not speaking from personal experience.)

The bat detectors soon start emitting a dry, crackling noise, as if someone is trying to speak to us from beyond the grave; it's the sound of the Daubenton's echolocation calls, Scott explains. Their social calls are likewise out of the range of human hearing, but on the sonograms they look like little rainbows. We see a couple of bats flickering in and out of the mouth of the tunnel. This is described as pre-emergence activity – the bats are testing the water and checking for any predators as they prepare to head out for the night. 'They get braver and braver,' Scott asserts, 'and then eventually one will just be like, "fuck it" and go.'

Sure enough, after a short while the crackling becomes almost constant as the bats come thick and fast. They whoosh out of the tunnel like early-morning rush hour, dark T-shaped blurs cutting through the beam of red light. I ask what kind of speed they can fly at; 'Seven metres a second!' replies Morgan, a more precise answer than I was expecting. As it turns out, this is one of the things that she and the team have been studying, through setting up torches 100 metres apart and timing how long each bat takes to travel from one to the other.

In a matter of days the bats will start swarming at the limestone caverns, their local hibernation site. In my head I imagine huge clouds of them, swirling like starlings in a murmuration, but Morgan quickly corrects me. 'They will aggregate and fly around in big cohorts,' she explains, 'but it's not like bees or wasps swarming. We don't actually know why they do it, although we think it's related to social bonds and mating.'

'It's an amazing thing to see,' Scott adds; 'I've experienced it once or twice, with fifty or more bats flying over my head at once.'

As they ready themselves to swarm, the bats are also mating. The males have sex on the brain, while the females are much more pragmatic, focused on gaining as much weight as possible and preparing their young for hibernation. Common to many bat species, Daubenton's use delayed fertilisation, so the females will retain the sperm inside them until they wake up in early spring. If there's enough food around, they'll soon become pregnant, often within just forty-eight hours. This explains why climate change is already playing havoc with bat populations – since they only have one pup a year, if they wake up in January during a warm spell the females can quickly become pregnant. Then, if the temperatures drop again and there isn't enough food to sustain them, they can end up miscarrying or reabsorbing the foetus.

One thing that bats do have in their favour is longevity – they are, proportionally speaking to their size, the longest-living mammal; a male Siberian Brandt's bat, ringed during adulthood, was found to be at least forty-one. In the wild, Daubenton's bats will live to about eight years old, but this is heavily skewed by the risk of predation and a relatively high mortality rate, especially amongst their young. 'If bats were the size of a blue whale, they'd outlive blue whales by quite some time,' Scott tells me.

As we've been chatting the time has flown by as swiftly as the Daubenton's. Tamar gives us the final tally – eighty-three in total – double what they usually record on a night like this. Perhaps the swarming has already begun, Morgan suggests.

Tamar heads to her car and Morgan and Scott walk me back to Dudley Station, for which I'm very grateful – although it's only a short walk my sense of direction is even worse than usual in the dark. We pass over a bridge alongside the train tracks and Morgan shows me the signalling lights that turn from red, to yellow, to double yellow, to green, indicating whether it's safe for trains to pass. She has recently found that noctule bats really dislike the yellow and will turn on their tails, switching direction as the lights change to avoid flying past them. She is busy gathering evidence to persuade her boss at Network Rail that the signals should be more bat-friendly.

Waiting on the platform for the next train, I am filled with admiration for Morgan and Scott; for their dedication and singularity of purpose as they seek to protect those creatures that lie sleeping in tunnels and man-made caverns while the human world rumbles on, largely oblivious to their existence.

GOING UNDERGROUND

It was autumn when I decided we needed a pond in our garden, following my afternoon with Emily from Froglife. Autumn isn't the best time of year for pond-building – the plants don't have time to establish before the weather starts to turn – but once I get an idea in my head there's no stopping me. I ordered a half wine barrel online and left it to fill with rainwater (which I had to repeat a couple of times to remove the strong scent of red wine). I piled logs around the barrel so anything that found its way in could easily find its way out. Over the winter the plants lay dormant, the pond empty. In springtime the water began to clear and I became borderline obsessed, checking daily for signs of life. For weeks there was nothing, aside from the occasional bloodworm.

Then, one day, all of a sudden the water was filled with squirmy things, metallic in colour, all hairy head and long tail, congregating around the inner sides of the barrel. I watched as an adult female delicately landed on the water's surface, brown wings folded, head and proboscis slightly raised as she dipped her stripy abdomen into the water to lay her eggs and produce yet more of the creatures.

Now, a few weeks later, I take out a small net and skim off the skins discarded by hundreds of offspring and empty them into the flower bed.

When people encounter *Culex pipiens* they will generally find a suitable implement – a newspaper, book, a tissue or sometimes simply the palm of their hand – and squash it. Otherwise known as the house mosquito, it has gained a bad reputation, and admittedly not without cause, although only the females will bite, and only in spring and summer. They are not fixated on mammals per se – they'll far sooner suck blood from a dove or pigeon, but a human will do if it's all that's on offer.

More than two decades ago, biologist Kate Byrne decided to dedicate her PhD to mosquitoes, planning to investigate whether insecticide resistance had been imported into the UK. But after a year she quickly learnt that it hadn't. 'I was left with all these *Culex* mosquitoes and I needed to find something else to do,' she explains over a Google Hangout. 'I had one small colony from a mosquito found on a Tube train at Elephant and Castle, so I decided to try and find out where they originated from.'

Kate managed to get an introduction to the Chief Executive of London Underground and he agreed to allow her access to the network for her research. Descending into the tunnels at night after the Tube was shut was a hazardous process – power cables and live electric rails had to be side-stepped and switched off; sumps, used to pump out water, had to be tested for methane and carbon monoxide. Kate climbed up and down ladders and dug through oil and muck.

After six months of work she collected six populations in total, including a cluster from Shepherd's Bush on the Central Line, where she got wind of reports that maintenance workers were getting bitten inside the tunnels. And the Victoria Line mosquitoes, which came from Finsbury Park, where she delved through filth beneath the track to find them. What Kate discovered as she studied their behaviour and genetics was nothing short of ground-breaking – she found that they were completely unique from other *Culex* mosquitoes.

The lifecycle of a surface-dwelling *Culex* mosquito is roughly as follows: in spring and summer the females lay their eggs in my tub pond (or any other available pool of water). It is during this period that they need the energy of a blood meal to breed and we are all at greatest risk of getting bitten.

Then, towards the end of the season when breeding is complete (here in London, around August time), the females begin to alter their behaviour, prompted by changing light levels. They shift to feeding on plant nectar, building up their reserves for the winter. As the season draws to a close, the males die off and the females go into hibernation, hiding in our log pile, perhaps, or the burrow where the bumblebees used to nest, or the dark corners of our shed. The whole cycle begins again the following year.

Very little of this applies to the London Underground mosquitoes. Their new home being a warm and stable environment, the mosquitoes have evolved to stay active and continue breeding all year round.

They've also adapted their feeding habits. With no available birds, they've turned to mammals – humans and rodents – in order to get their blood meal, although most of the time they no longer need access to blood at all. 'Because humans are so messy and disgusting, the water in the Underground tunnels or around the tracks where the mosquitoes lay their eggs contains lots of nutrients,' Kate explains, 'so much, in fact, that the females don't need to bite at all in order to lay their eggs.' Kate found that she could keep her Underground mosquitoes going in the lab for generations before having to supply them with blood. 'When the lines started to get a bit weak, because they hadn't had blood for generations, I would let the females out one at a time to have a little feed, and that was the only time I got bitten,' she casually adds.

There are clear genetic differences too. If you compare surface-dwelling *Culex* mosquitoes, you'll find they are genetically variable, but Kate discovered that the same isn't true of the Underground mosquitoes. This suggests that the Underground populations originated from just a few brave individuals which ventured down from the surface, travelling via the trains, perhaps, or funnelling through the air vents.

These changes have taken place over an astonishingly brief period of evolutionary time, bearing in mind the age of the London Underground. But whether or not the mosquitoes have now become a unique species is more of a hot topic. One crucial determining factor is their willingness to interbreed with their surface-dwelling ancestors. 'I couldn't cross any of mine with the surface ones,' Kate tells me, 'but the surface mosquitoes won't breed in a laboratory environment anyway. They don't generally crossbreed, but it's not impossible.'

Then again, even if they did, it wouldn't necessarily mean they were one and the same species. 'It's a bit like the difference between a Siberian and a Bengal tiger,' Kate explains. I reference zee-donks (zebras crossed with donkeys) and ligers (lions with tigers) and mules. 'Exactly,' she replies; 'the level of success and what you get from a cross depends on how long the two species have been isolated from one another.' In the case of the Underground mosquitoes, she concludes, 'we would probably say that a speciation event has occurred'. In recognition of this, the mosquitoes have been given their own name, *Culex molestus*, in reference to their human-biting ways.

Not everyone welcomed Kate's findings – the study drew a surprising amount of negative attention from religious fundamentalists, who wasted their time trying to discredit it.

And as if the presence of a new urban mosquito wasn't enough for people to get their heads around, Kate also made another strange discovery: that each individual population of *Culex molestus* was genetically unique to their Tube line. Kate concluded that every time an Underground colonisation event occurred, the mosquitoes began to adapt, restricted to their Tube line by the movement of the trains and the way that the tunnels are ventilated. This allowed her to distinguish between a Central Line and a Bakerloo Line mosquito. Kate found that the mosquitoes she collected from Liverpool Street were genetically more similar to those from way up the line in Shepherd's Bush, 6.5 miles away, than her mosquitoes from Waterloo, just 2 miles in distance but on a different line.

Not content with watching the plain old *Culex pipiens* buzzing about in my garden, as soon as I learnt of the London Underground mosquito I sought to track it down for myself. First I emailed TfL's press office, then I spoke to a friend's dad who worked for the London Underground. I even tried the Deputy Mayor of London for Transport, whom Ben used to work for. 'Let me see what I can do,' she replied, adding: 'I don't like the sound of the uniquely developed TfL mosquito!!!' I finally heard back from an otherwise helpful and cheery member of TfL's PR department, who informed me that the mosquitoes weren't something TfL was looking to promote.

As it happens, Kate explains, even if I had gained access with the maintenance crews at night, my chances of seeing one were from low to zero – like a needle in over 100 miles of haystack. Even since she conducted her study the mosquitoes have become increasingly rare as the water table has been lowered, leaving fewer underground pools in which they can breed.

And TfL needn't worry – *Culex molestus* is harmless.

While it's true that their tropical counterparts can spread dengue and elephantiasis, we don't have those diseases in this country, so even on rare occasions when they do bite humans the worst that will happen is a slightly itchy bump to the skin. 'You need to keep your parasites and your viruses at bay to keep the mosquitoes safe,' Kate observes.

After generating a flurry of global attention Kate moved on from mosquitoes to focus on grazing mammals, her projects including conservation of rare sheep breeds and pair-bonding in donkeys. She now lives in Edinburgh (hence why we're meeting virtually), where there's no Underground or Metro system, she adds slightly wistfully; instead of harbouring mosquitoes she devotes her time to keeping horses.

Watching the latest batch of larvae forming a scummy rim around my tub pond, I feel a slight twinge of sadness that the one and only urban adapted London Underground mosquito could be facing extinction. I wonder what would become of my *Culex* mosquitoes if they were to find their way down the road to King's Cross – whether they would board the Victoria Line, or start a new lineage on the Northern. Or perhaps they'd head to Heathrow on the Piccadilly. The odds might be extremely low, but if you do see a mosquito on the London Underground – inside a train carriage, or resting on one of the adverts plastered to the Tube walls – there's always a chance it might be a *Culex molestus*. Or perhaps it's just at the start of an evolutionary journey towards becoming one.

INTO THE SEWERS

In common with mosquitoes, there are few urban creatures more maligned than the rat. 'You're only X number of metres from a rat,' the urban myth goes, with the number varying each time, depending on who is relaying the information.

I first learnt that I was attracting rats to our garden on a sunny April morning, when I attached a small camera to the bird feeder. A minute or so into the footage a pointy snout appeared, and then two small hands, as the rat pulled itself up to the rim of the seed dish. It had glossy, sandy-coloured fur intermixed with darker streaks, its nose and whiskers twitching as it crunched on the grains. It stayed only a few seconds – daylight is a risky time for a rat – before disappearing off the edge, its scaly tail, a rat's balancing mechanism and the source of so much revulsion, trailing after.

I named it Ratticus Finch and shared the video with some of my neighbours, most of whom were unimpressed, although one conceded it was 'quite cute'. A few weeks later, our neighbour Alex messaged: 'Garden update: Ratticus Finch found rigor mortis in my garden. It seems that some cat injured him fatally.'

'Hope he doesn't leave a grieving wife and family behind,' our neighbour Matt replied.

The next morning, I was relieved to find Ratticus alive and well, back at my bird feeder, and beside her (I concluded it was a her, male rats being pretty easy to identify from behind) a Ratticus Junior. Although I haven't seen any of their kin for a while, I'm sure the rats are still around – in the evenings our dog Fox senses their presence, whirling around in circles as she charges into the bushes in the corner of our garden that's next to an alleyway.

My view on rats has surely been shaped by the experience of keeping pet 'fancy rats' as a child (I imagine someone added the 'fancy' to improve their public image). One in particular, a dark-brown-coated individual we called Bruce, had a lovely temperament and would sit on my mother's shoulder while she was doing the ironing, making a gentle chattering sound with his teeth, a sign of rattish contentment. He was so tame that we could set him down at one end of our long garden and he would run all the way back to us, clambering on to our feet and squeaking to be picked up. Bruce was also incredibly fastidious, spending a large proportion of his time grooming himself with his pink paws.

But wild rats are very different to pet ones, as Bobby Corrigan, one of the world's few urban rodentologists, gently explains to me over Zoom. Bobby is a warm and thoughtful New Yorker who cares enough about rats and mice to have devoted his life to their study; he even takes people out on 'rat safaris' in the city where 'pizza rat' – a rat dragging a slice of pizza down the stairs of the New York Subway – became a viral sensation. In particular, he is an expert in the conflicts of living alongside rats. 'These days we are all virus-conscious,' he observes; 'we've all seen how a virus has changed the world and killed hundreds of thousands of people. Well, if a rat comes into your home, that animal has been living in the wild, with a capital W. It has been eating other wild things. It has been getting its feet and its tail dragged through faecal material and blood and death. And those are the origins of dangerous viruses and bacteria.'

It's true – in the UK, though rare, we still have cases of Weil's disease, a form of bacterial infection which is caused by coming into contact with the urine of infected rats (or cattle), usually via contaminated waterbodies. The bacteria make their way in through an open wound and cause flu-like symptoms which, in the most severe cases, can be followed by kidney, liver, respiratory and heart failure.

Most people's fear of rodents – and indeed extreme revulsion towards them – is thus deep-rooted; as Bobby asserts, we are hardwired to be wary of spiders, mosquitoes, snakes and things with snake-like tails. Our species has lived through plagues and pandemics and we've learnt over the course of millennia, from sheltering in caves and then yurts and huts and eventually houses and flats, that anything slithering along the ground or scuttling about at night and eating our food while we sleep has the potential to harm or even kill us.

Like mice, rats have adapted alongside us. Bobby likens them to pilot fish, which follow whales and sharks across the ocean as rats have followed humans, feeding off their leftover food and parasites. Rats don't hibernate, nor are they able to migrate when the weather grows cold and resources are scarce. 'They've seen us,' Bobby explains in his soft New York accent, 'and realised, "Hey, these animals, human beings, that are living in these caves, keeping them warm, leaving out scraps of food – why don't we just move in with them?"'

Rodents comprise the largest order of mammals, with 2,300 species. Humans, by comparison, are of the order of primates, with between 190 and 448 species, depending on the classification used. There are just six species of great ape: gorillas, chimpanzees, bonobos, Bornean and Sumatran orang-utans and humans, while there are around sixty species of *Rattus*, the genus to which rats belong. There are, however, only two species of urban rat – the black rat, *Rattus rattus*, also known as the roof or ship rat, and the brown rat, *Rattus norvegicus*, the Norwegian rat.

Despite their name, black rats are often a grey-brown in colour, with lighter bellies and dark tails. Through the presence of urban rats we can trace the movement of global trade – black rats first reached Britain on trade ships in Roman times, having originated in South East Asia, before arriving in the new world in the sixteenth century, carried on ships from Europe. Although still common in other parts of the world, they are now one of Britain's rarest mammals, driven out with the decline of industry when their final outposts, the dockyards, were modernised.

Black rats have long been blamed for transporting the fleas that brought the deadly bubonic plague to Europe, killing as many as half of its peoples within just seven years, but some scientists have recently begun to dispute this theory, asserting that the lack of rat remains from that period and the speed of the spread suggest that the infected fleas and lice spread mainly from person to person.

Brown, or Norwegian rats, which have since become dominant, are larger and generally lighter-coloured. Also incongruously named, they don't come from Norway at all and probably originated in Mongolia. 'We always like to blame you guys, the British, for our rats,' Bobby smiles, 'but you got your rats through trade from the East and so on and so forth.' The brown rats hitchhiked along the

Silk Road on wagons filled with silks, spices and new technologies like paper and gunpowder.

The other difference between black and brown rats is their orientation: one goes up and the other goes down. Brown rats are geotaxis-positive – their instinct is to burrow. Black rats are geotaxis-negative – their instinct is to climb. But since they are both opportunists, adaptability being key to their success, this rule often proves pretty meaningless, especially in the urban world. It does, however, explain why a brown rat in the city, whose instinct is to construct a simple burrow a half-metre into the earth or to find shelter in a hollow log, is drawn towards our foundations, beneath our decking and deep down into our drains and sewers.

To help me better understand the life of an urban rat, Bobby puts me in touch with Manchester-based Davy Brown, Managing Director of RatDetection.com. Davy works with pest controllers, using drain surveys to locate unwelcome rats at source and prevent them from entering our buildings in the first place. After meeting at a conference, Bobby arranged for Davy to present his work to the local authority in New York, where their only reaction to rats was to reach for the poison, while no one was looking seriously at the source of the problem – the drains. 'Crazy, in't it?' Davy observes.

I meet Davy at the site of a fast-food restaurant in a town centre in Buckinghamshire in south-east England; I can't be more specific than that, for reasons that will soon become apparent.

Davy is a cheerful, straight-talking Mancunian who has been very accommodating towards my rather unusual request that we meet at the site of a rat infestation. 'You need to get out more, Florence,' he jokes. He has been called in by Stuart, a local pest controller, to help figure out how the rats are getting in. The building was once an old pub and probably has no real foundations, immediately making it higher-risk, Stuart explains.

Rats have an irresistible urge to chew, although no one is quite sure why. It's true that their incisors continue to grow at 11–14 centimetres per year throughout their lives, but the long-held belief that obsessive gnawing is required for rats to wear down their teeth is erroneous – this happens naturally. What is for certain is that those teeth are incredibly powerful – stronger than aluminium, copper, lead and iron and comparable to steel.

At this site the rats have already begun to munch through some of the cabling, causing electrical shorts. This can have serious consequences – depending on which study you look at, between 5 and 25 per cent of unexplained house fires are thought to have been caused by rodents.

Often, Davy explains, when people find rats in their property it's up in the roof or the loft, so they don't even bother to check the basement, cellar or drains. But as well as being accomplished diggers, rats are excellent climbers (like a rat up a drainpipe – you would think we'd remember the expression) and can tunnel their way into cavity walls to gain the run of the building. They can also collapse their skeletons to squeeze through the most improbably tiny of gaps, down to the size of a fifty-pence piece.

As the first phase of their detective work, Davy and Stuart have removed a manhole cover in the restaurant's adjoining yard and stuck what looks like a long black hose with a camera on the end into the drain. The pair only arrived half an hour or so before I did, but they've already found the source of the problem, which Davy shows me on his monitor – a huge, corroded hole in one of the cast-iron pipes. The only thing that he's slightly perplexed by is that it's unusual to find pipes like this outside. He suggests we go into the cellar to see if we can find any more clues down there.

The cellar is accessed through a hatch in the kitchen and down a set of steep stone steps, which we carefully descend while bowing our heads. It's cool and dusty; dense mats of ancient cobwebs coat the ceiling and crumbling brickwork and old beer-bottle tops litter the floor from the time when this was a pub. Along the edges of the walls are small plastic trays of blue rodenticide, the lurid colour designed to deter humans from consuming it. I'm slightly relieved to find that the trays appear untouched: most rats are naturally 'neophobic' – wary of new things in their habitat – and will treat poison with suspicion if they can find an alternative source of food. Stuart is quick to explain that he uses poison sparingly and only in places where he knows there's minimal risk of another creature ingesting it, either directly or indirectly through eating a poisoned rat. One of the biggest issues, he explains, is that it's now astonishingly easy for anyone to get their hands on rodenticide, with most members of the general public having little understanding of the harmful consequences that come from improper or ill-informed use (beyond the intended harm to the rats, of course). 'I don't like killing anything myself,' he adds with a note of regret. 'If you look in my van, I have far more builders' stuff than I have poison.'

We don't go any more deeply into the subject of rodenticides, but the truth is that the most common type of poisons used are anticoagulants which, when ingested, cause the rat to bleed internally. It often takes several poisoned meals to finish a rat off, and rats will generally return to their nests to die a painful death. In public spaces the poison is often housed in opaque plastic shoebox-sized boxes – once you start spotting them, you'll realise that they're everywhere; I found one by a set of traffic lights on busy Camden Road the other day.

It's a thorny issue and I do know from my conversation with Bobby, who unsurprisingly is no fan of poison, that none of the options are particularly palatable. Live trapping isn't much better, he tells me: releasing a rodent in unfamiliar territory, though it may assuage some guilt, often results in the animal slowly starving to death. The old-fashioned spring-loaded traps are generally thought to be the most 'humane' route, when placed in boxes that steer the rodents into position and break their backs, which should at least end their short lives swiftly.

Back to the job in hand, we can only have been in the cellar a couple of minutes before Davy asks Stuart, 'Is that a cast-iron pipe you're leaning on? Is that the hole?' There is indeed a huge old pipe, brown and corroded, running across the cellar at chest height. At one end there's a sizeable chunk missing from the top. 'There's your problem my friend!' Davy proclaims triumphantly. It is his suspicion, he adds, that the pub used to have an outdoor toilet, and the redundant pipework was left behind when it was removed. For a rat it couldn't be easier to pull itself out of the hole and gain access to the building, while maintaining a clear route back to the sewers.

In theory the sewer itself should provide a rat with everything that it needs – warmth, shelter, water, food. Rats are both omnivores and true generalists – there isn't much they won't eat. There's the food that we wash off our plates, the discarded fat from the restaurant trade, the bugs swept down with the surface water, the undigested food in human excrement – something to suit all palates and tastes. There's a distinct lack of predators too. 'Sewer rats don't know about birds, badgers, foxes, dogs, cats,' Davy observes; 'when people ring up and say there's a rat walking across my patio in broad daylight, it's probably a sewer rat, because they don't know the risks. Either that, or it's a rat that's had rodenticide, because rodenticide sends them a bit drunk and they get desperate.'

He shows me a video filmed during one of his surveys, in which he inserts his camera device into the pitch-black drain. Undeterred by this strange new object and the light it emits, a young-looking rat bowls up and fearlessly licks the lens.

The problem isn't the rats that spend their whole lives in the sewers, where the population size is determined by the amount of available food and suitable nesting space – in fact, allowing a stable population to establish itself is often the most biologically sound answer to species-based conflicts. Sewer rat numbers may well also decline over the next few decades as climate change causes more heavy downpours and the sewers are regularly flooded. 'It's only when they pop their heads out to feed off the table of man that it becomes an issue,' Davy observes, 'and the trouble is, if sewers are the motorways then private drains are the A roads.' If there's a way out of the sewers and into our buildings, rats will certainly find it – they can hold their breath for a number of minutes and swim along for hours, doggy-paddling or floating through the darkness. Davy even claims that on occasion they've been known to swim up a toilet U-bend, 'So blokes – always put the seat down,' he jokes.

If one rat finds its way out, you can guarantee others will too – not least because they leave a pheromone trail behind them. Davy once visited a residential property in Guildford, where the pest controller who called for his help had pulled seventy rats out of the loft in eight weeks; while Davy was surveying he came down with another six. Davy discovered a defect in the drains leading down to the sewer, and at the same time he found that the main sewer was blocked. This caused the water to rise, forcing every single rat within 300 metres to move upstream, swimming for their lives. They all escaped into this house, using the plumbing tree to reach the attic where they soon began to breed – something that rats are very adept at. In the wilderness they create nests out of bits of plant matter or blades of grass, but in an urban environment they'll use plastic bags, cloth, foam, loft insulation – anything they can get hold of and shred. Males and females can mate up to twenty times a day, and with a gestation period of twenty-one days the females are able to produce five litters of up to fourteen pups a year, given half a chance.

In the case of people's homes, the first point of access is often garden decking. 'Decking is the devil's work,' Davy complains, 'we call it the rat penthouse. It lets them move about unseen next to your foundations.' Ground-floor extensions are another key access point, where kitchens have been moved, leaving the old drains beneath. Once again, the fault lies with us humans and the way in which we build – old pipes blocked up with half a brick or a bag of cement and metal materials replaced with cheap plastic. 'Who knew rats could chew through plastic? Everyone!' Stuart quips.

One thing you can be fairly sure of is that if there are rats in a building, there won't be any mice and vice versa; mice being the smaller species, rats will quickly predate them. In buildings that are poorly constructed, if a rat population is destroyed without tackling the entry points, an explosion of mice may follow. Mice can tell by the smell of the previous rat colony's excrement whether or not they're still active. They won't let prime real estate go to waste – where they determine the rats have vacated, they'll quickly take up residence.

Mystery solved, Davy, Stuart and I clamber back out of the cellar. The solution is clear – cover the hole with something the rats can't chew through until the pipe can be replaced. While Stuart goes to his van to find the right tools, Davy tells me about some of his recent clients. One, who works in a high-powered job, twice rang the office to ask for help with a rat problem in his home. Both times he broke down, unable to finish the conversation because he was so distressed; he was ready to sell the house, he said, just to escape the rats. This is illegal, though not uncommon – over half of infestations are thought to have been inherited. People get desperate, Davy explains, when they call in the so-called experts, only for the rats to return soon after.

Meanwhile, Davy continues, he recently went to see a woman well into her eighties whose son had returned from overseas and was horrified to find she had rats in her house, but she wasn't bothered at all. 'She had a high-level grill,' he tells me, 'and she said, "Every morning I put me bacon under the grill and he sits over there on the counter top, next to the sugar bowl, watching me." The rats had perceived that she wasn't a threat and, bold as brass, would wait for her leftover bacon.' Davy adds that he politely declined her offer of a cup of tea.

Stuart returns from his van clutching a large roll of black adhesive tape with a core of wire mesh that even the rats can't munch their way through; or that's the theory at least – both he and Davy freely admit that the rats are often one step ahead. 'They never stop learning – born survivors – they'll be here long after we're gone!' Davy observes with a certain reverence; 'I work with thousands of pest controllers and I don't like rats being killed. I'm a fan of rats – they pay me mortgage.'

If there's one thing that Davy's work shows, it's that we are often too quick to reach for the poison, instead of taking the time to understand those creatures that we call pests and modifying our own behaviour. But this requires acknowledging that they are worthy of our attention in the first place, which, sadly, most people are unwilling to do.

This takes me back to my conversation with Bobby in New York, who describes how decades ago, when he was working on his PhD, his peers flew off to far-flung locations to study whales and wolves and tigers and other exotic species. Meanwhile, Bobby moved into a nearby barn with a colony of rats. During that time, the thing that surprised him the most was the variety and complexity of behaviour that he witnessed. When we use the word 'rat' to describe other human beings we associate it with the worst of all behaviours – a rat is devoid of morals, loyalty or scruples; someone who deserts those around them without a second thought during times of trouble. But the irony is, Bobby admits, that in many ways the behaviour he witnessed in the barn resembled our own: 'I saw aggression, I saw companionship, I saw joy, I saw empathy and altruism; who would've thunk it, but rats will help other rats without expecting anything in return.' This empathy is all the more remarkable when you consider the competitive world that rats live in. Marking the rats with numbers on their rumps, he came to realise that they had distinct personalities. 'Anytime I saw number four, I was like uh-oh, we've got trouble, and there he would be with the other males duking it out. But then when I saw number 12, I knew that this is a good rat, this is a kind rat, this is a wonderful rat!'

During Bobby's time in the barn the rats never showed any aggression towards him. Gradually they became accustomed to his presence and would cautiously approach for a treat. He became accustomed to them too; lying on the floor in a sleeping bag he would sleep soundly amongst them. When it came time to leave, he did so with a heavy heart. The farmer who owned the barn had given him sole use of it for the duration of his research, but that was as far as he was willing to go. 'I just remember saying, I understand, and I got in my truck and drove away,' Bobby reflects; 'I lived with them and they got to know me, I got to know them. And in the end, after befriending them, I turned them over to be annihilated. I think about that a lot.'

We despise and punish rats for their success, driven by our fear of abundance – of swarms and surges, engulfing what we have built for ourselves – a form of nature that we cannot control (though we call it 'pest control' in a bid to assert our dominance). We condemn an entire species on moral grounds for their strength in numbers – another irony, considering our own impact on the planet – and at any rate, nature was never, and never will be, a moral force. We show little gratitude towards a mammal that is responsible for so many of the advances in medical science that we've achieved through subjecting them to lab-based experimentation for over 200 years. By definition, a pest is like a weed – an animal out of place. And yet so often we humans are the ones inviting them in. Rather than reaching for the poison (as we reach for the herbicide), we could start by designing our buildings better and correctly disposing of our waste. Rats are the ultimate urban exploiters, and as a result they find themselves at the heart of the conflict between humans and other species. Perhaps it's time we showed them some respect.

Tailpiece

*Even in the most sordid street the coming of spring
will register itself by some sign or other, if it is only
a brighter blue between the chimney pots or the
vivid green of an elder sprouting on a blitzed site.
Indeed it is remarkable how Nature goes on existing
unofficially, as it were, in the very heart of London.*

George Orwell, *Some Thoughts on the Common Toad*

'All corners of the city have stories of nature to tell,'
reflects the London Wildlife Trust's Director of Conservation, Mathew Frith, as we stroll through Victoria Park
in east London on an overcast mid-September afternoon.

Mathew, who has light-grey hair, a hoop in one ear and
big silver rings on his fingers, has been involved with the
Trust since not long after its inception over three decades
ago and is its longest serving member of staff. One of
just a handful of voices, including that of green-roof
pioneer Dusty Gedge, with whom he fought to protect
the biodiversity of Deptford Creek, Mathew has been
speaking up for urban wildlife since the days when few
others would listen. If anyone is in a position to tell these
stories, it's him.

In the 1970s and 1980s urban nature was generally
viewed as being of little ecological value and no strategic
interest. 'It's a bit of a flippant view,' Mathew admits, 'but

the reason the Trust was formed was because nobody else was paying any interest.' Back then, there were just 6.5 million people living in the capital. Now, at nearly 9 million, London is close to becoming a megacity, according to the UN's threshold. While we may not have the vast slums or favelas of megacities such as Mumbai and Rio de Janeiro, virtually all of our urban hotspots suffer from a shortage of adequate housing, with overcrowding and homelessness posing huge problems: families confined to one-bedroom flats in high-rises and tenement blocks, children growing up in a world of concrete and tarmac, with little connection to nature at all. Few species can survive in this exclusively rocky landscape without soil or vegetation; even fewer can thrive, and that goes for us humans too.

This puts added pressure on our existing green spaces. When it was decided that HS2, the new high-speed railway designed to link London, Birmingham, Manchester and Leeds, would run through my borough of Camden, our councillors were left with some impossible decisions to make. Should they protect the precious few green spaces left in the densely populated, high-deprivation wards that were to be the most affected, or build new housing for those whose council-owned homes were about to be demolished? Should they allow these people to stay within their local community, or deprive even more of their access to nature?

As is the case with so many cities, this access isn't equally distributed across the capital. In the south-west there are the sprawling deer parks of Richmond and Bushy. There's the River Thames itself, cutting through the city; there's Hampstead Heath and Highgate Woods in the north-west. But out here in the east such historic landscapes were largely obliterated – through agriculture and then, following large-scale bomb damage, as a result of post-war development and land clearance.

In 1984 the London Wildlife Trust began to survey the whole of London to identify the most important spaces where nature was succeeding in the capital in order to give them a level of protection, and to help ensure people were able to access those spaces to connect with nature. The Trust, working with the then Greater London Council, developed and promoted the concept of a SINC – a Site of Importance for Nature Conservation. While this didn't afford the same level of protection as a SSSI, it was a more appropriate tool for an urban context, where our green and blue spaces, though extremely valuable, are less likely to support the rarest, most specialist species of flora and fauna, and where development is a critical pressure. Today, London has thirty-seven SSSIs, more than triple the number just a few decades ago, but in contrast there are now over 1,600 SINCs, giving at least some protection against the ever-looming threat of development. The remaining challenge is to monitor our existing SINCs and gather data to build a case for new ones to be designated – a task that often falls to local volunteers and community organisations.

In establishing the SINCs, the data has since been used to identify and target areas of deficiency that are more than a kilometre away from the nearest 'high value' SINC. There are still around 33,000 hectares of London that fall within this category, and unsurprisingly there's a strong correlation with multiple indices of deprivation. Similarly, a 2020 survey of 2,000 people from across the UK by the walking charity The Ramblers found that wealthy white people are far more likely to have green or blue spaces on their doorsteps than those from deprived or ethnic-minority backgrounds.

Though Victoria Park might not be a haven for rare species – with its large expanse of short-cropped grass and modest number of trees, it certainly wouldn't qualify as a SSSI – it remains one of the only sizeable stretches of green in the surrounding area. Known locally as the People's Park, it was first opened to the public in 1845, the land having been purchased by the Crown Estate following a recommendation by epidemiologist William Farr and a mass petition to the Queen. It became an essential amenity for the working classes of the East End and a hub for political meetings and rallies. Today its old concrete swimming pond and boating lakes attract dragonflies, damselflies and increasing numbers of waterbirds, probably also drawn in by the two canals that bound the park – the Regent's Canal to the west and the Hertford Union Canal along the southern edge.

So the park has huge local importance. It is therefore of little wonder that there was an outcry when, during the first lockdown of the coronavirus pandemic in spring 2020, the park was shut by the council, leaving large numbers of people with nowhere else to go to take their exercise or escape the cramped conditions of their homes.

One of those people was my then heavily pregnant friend Emily, who lives in a flat near the park and was left with only a short section of canal along which she could take her walks. She took solace from a pair of nesting coots, those highly successful members of the rail family, with sleek black feathers, beady red eyes, stomping blue feet and strange lobes of skin on their toes. Rails are generally known for being shy birds, but urban coots are quite the opposite – I was once sitting innocently by the Regent's Canal when a coot marched up and repeatedly jabbed at my shoe with its beak; perhaps I looked at it the wrong way, or committed some other misdemeanour for which it determined I must be punished.

Emily watched as the coots continued haphazardly to build their nest, intermixed with crisp packets and bits of old rubbish. She patiently waited for the mottled-blue eggs to hatch, giving her friends and family progress updates. Over Easter she painted a wooden egg for her partner, with an image of the coots on one side and her unborn child on the other. Finally the chicks hatched, but only a few days later she returned to find a drowned black creature floating near the nest. Upon closer inspection, discovering that it was wearing a collar and was larger and furrier than the coot parents, she concluded that it was someone's unfortunate pet. On her final visit the cat was gone, but two of the chicks had perished, their scruffy black bodies and red and yellow tufted heads ruthlessly woven into the nest itself as their parents continued to build on top. 'So it turns out I should definitely have researched coots in more depth. Baby-killing is entirely normal family behaviour it seems,' she messaged mournfully. It's true – coots can be outstandingly vicious, the parents sometimes pecking or shaking their own chicks to death when they beg for food, especially when there isn't enough to go round.

It's easy to romanticise nature, especially when we're sharing such close quarters, but just because it's adapting to our cities doesn't mean that it will conform to our per-ceived standards of behaviour, as we have discovered time and time again throughout this book. Mathew points to the magpies rooting around on the ground by a nearby bin. 'They're one of my favourite birds,' he tells me, 'but people generally seem to have a problem with them, because they are very good at pulling out all the fledglings in the spring.' To a magpie, a baby robin, blue tit or sparrow is simply a nutritious, high-protein meal with which they can feed their own young. But those with a strangely hierarchical view of how nature should be believe that we should intervene – there's even a campaign group that calls for the persecution of magpies and other members of the highly intelligent crow family, along with raptors like sparrowhawks too, under the guise of protecting our songbirds.

It's a similar story with the ring-necked parakeets that dominate Victoria Park and are now spreading beyond London and into south-east England; there are still wide-spread calls for them to be culled, nonsensical though this is. 'The horse was out of the stable door back in about

1996,' Mathew declares, arguing that these lurid-green immigrants are the perfect expression of our multicultural city, even down to the urban legends about how they first got here. Some people claim that the parakeets are direct descendants of a pair released by Jimi Hendrix on Carnaby Street in 1968; others that they escaped from aviaries during the Great Storm of 1987. It has even been alleged that they arrived in 1951 on the set of the film *The African Queen* starring Humphrey Bogart and Katharine Hepburn. (Recent analysis suggests that the population has established and grown through a number of small-scale pet releases, both accidental and intentional.)

'A couple of years ago we were taking some people on a walk and we saw one fly up into a hole in an oak tree, yank out two baby squirrels and break their necks,' Mathew confesses, 'and everyone was really conflicted, but I still think they're fantastic.'

The nature of Mathew's role forces him to face up to these sort of conflicts, which are only heightened in an urban environment. This includes weighing up the different schools of thought on how we should be managing wildlife in our cities. There is a pervading view that nature should be in some kind of balance, but in truth it is a dynamic system, forever changing and evolving. 'There are reasons why we have larger numbers of crows, foxes and magpies than we do in the countryside,' Mathew asserts; 'people presume nature is out of kilter, but the truth is that you've got certain species that become extremely well adapted to human habitation; they thrive because of what we do to our urban spaces, the food we leave lying around, the lack of predators, so they are clearly going to exist in greater numbers in our cities.'

Confronted with such complexities, traditional approaches to nature conservation can hit a bit of a brick wall (both literally and metaphorically) when it comes to urban wildlife. The prevailing assumptions about how nature is going to behave go out of the window when you're contending with extreme human intervention, disturbed and eutrophic soils, the urban heat island effect, high levels of air pollution, foreign species that have been introduced over millennia, either intentionally or unintentionally, and ecosystems that we've engineered almost entirely ourselves. 'In places like Victoria Park,' Mathew concludes, 'nature behaves differently.'

Leaving the park, we head towards Wick Woodland, which Mathew is keen to show me as a model for urban woodland management done well. 'We're very bad in cities at letting go,' he explains, telling me the story of a woodland he managed himself more than thirty years ago. 'Best job I ever had,' he claims, 'but through my training there was a kind of methodology that everyone adopted in terms of how you manage the woodland. So I piled in and followed that methodology for the first couple of years. It was only over time that I started listening to the wood, watching the wood and recognising what it was trying to do, and then using that to shape my interventions.'

Upon arrival, we find that the entrance to the woodland has been blocked. We pause for a moment, considering our next move, before quickly climbing over the 'No Entry' sign.

We soon discover that a burst Thames Water pipe has flooded the entire area. The wood smells of damp, and newly formed streams trickle down the many paths. The ground is strewn with dead worms, drawn to the surface by the water, where they then perished, their bodies white and puffy. Mathew semi-jokingly suggests that we head for the high ground. Climbing over a patch of stinging nettles, we eventually find a track that has escaped the worst of the flooding. These woods are generally well used, not just in the daytime but all through the night as millennials and gen Z-ers spill out of their warehouses in Hackney Wick, holding unofficial raves through to the early hours, until they're interrupted, bleary-eyed, by the early morning dog walkers. But today, thanks to the less than ideal conditions, we have the woods to ourselves.

Wick Woodland represents one of the best examples of modern tree-planting in London, Mathew asserts. Formerly a derelict piece of land, in the early 1990s it was planted up with predominantly birch and willow. The white papery bark of the slender silver birches gives the wood a mystical quality, the branches melting into the overexposed grey city sky. Silver birch is one of the first species to naturally colonise patches of bare land, supporting a wide range of insects and other invertebrates. Its leaves create an airy canopy through which the light can easily penetrate, allowing plenty of plants to grow beneath – on either side of the paths a mass of brambles, ivy and scrubby vegetation has been recently cut back in preparation for the winter.

One of the key factors that has led to Wick Woodland's success, Mathew explains, is that it is responsibly managed by a volunteer group known as the Tree Musketeers, who plant and care for trees across Hackney. Working with local residents, the council and other organisations, the group regularly holds workshops to teach forestry skills to as many people as possible. Urban conservation is at its most successful when it meaningfully engages and involves the local community.

In writing this book I've encountered dozens of such volunteer groups, many of whom have taken matters into their own hands. This includes the award-winning Wildlife Gardeners of Haggerston, founded by local couple Gideon Corby and Esther Adelman, also based in east London. When Esther and Gideon moved into their new-build flat overlooking the Kingsland Basin, they discovered that as part of the redevelopment of the basin – which is now flanked by buildings on all sides – their housing association had agreed to include a number of ecological enhancements that never materialised.

After years of fighting and guerrilla gardening, they secured the necessary funding to officially transform the basin. Together with a group of fellow residents they have truly brought it to life, along with a few hundred metres of the adjoining Regent's Canal. The water is now home to an archipelago of floating islands, much like those that were installed by Paul and his volunteers in Birmingham, awash with reeds, yellow flag irises and purple-loosestrife, water mint and herb-Robert. Against the steel-and-glass backdrop of London's looming Square Mile, they've filled tree pits with native flowers, along with crowd-pleasing hollyhocks. Using a de-paving grant from the Mayor of London, they've removed sections of paving to make way for more plants. And they've fought to prevent the sprigs of common toadflax sprouting from the side of the pavement from falling victim to the council's herbicides, preserving its yellow flowers like tiny snapdragons for visiting buff- and white-tailed bumblebees.

Then there are the volunteers at the Gartnavel Royal Hospital in Glasgow, a mental-health facility providing in-patient psychiatric care, who have transformed a patch of the hospital grounds into a garden for patients, the community and wildlife. Forest Gardener Camilo Brokaw has led the charge, with thoughtful layered planting that weaves together as many species as possible. This includes an array of fruit and vegetables to reconnect local people with their food at source, as well as dead hedges and bug hotels to encourage invertebrates and birds.

Since Camilo took over stewardship of the garden, he has focused on transforming prevailing attitudes towards weeding and mowing, with a view to creating a self-perpetuating habitat in which he and the volunteers can play a role, without being the dominant force. He's now looking to expand further into the extensive, and currently underused, hospital grounds to create corridors that connect the garden with other pockets of biodiversity.

These are just a couple of examples of countless people across the country who are transforming their cities to let wildlife in. 'I can't believe where we are today,' Mathew opines, 'with the level of consciousness and concern. When I first stumbled into this job, we didn't even have the planning protection for our own sites that we have now. It's true that we're facing massive declines in certain species and damage to certain habitats, but we've also got more tools and more awareness than ever before.'

Balancing the needs of people with those of nature is never going to be easy, especially where we're living in such high densities. When we talk about access to nature we also have to acknowledge that the more we humans use our green spaces, the more pressure we put on the wildlife that relies on those spaces to survive. Fortunately, most urban biodiversity has become fairly tolerant to high usage, so long as we treat it respectfully. But there's a wider point to be made here too, because if people don't have access to nature, they simply won't know that it exists in our cities in the first place, and they won't have a reason to protect it.

'I'm as concerned about the future of our tigers and elephants and black rhinos as anybody,' Mathew reflects, 'but conservation starts on the doorstep and until we know what we've got here where we live, and we're willing to work with that, we can't be telling anybody else what to do.'

'We might not like to admit it, but nature conservation is a political act,' he continues, 'because we're making decisions about how land is used, who it's used for and how it's managed. Using that term causes tremors amongst ecologists, because they argue that what we do is science. Yes, it's informed by science and how we understand nature, species and their habitats absolutely has to be grounded in science. But how you then deliver it – that has to be through a prism of culture, politics and community.'

Seven months later, on a disappointingly wet and chilly day in mid-May I meet with Mathew again; this time in Camden Gardens, just around the corner from where I live. I've asked him to join me on a walk along the proposed route of the Camden Highline, a prospective 'park in the sky' which, if approved, will run all the way from this small public garden to the King's Cross end of busy York Way, just five minutes' walk from the station. We're joined by Simon Pitkeathley, Chief Executive of the Camden Highline charity, which hopes to secure the necessary funding for the project to go ahead.

The Highline will make use of the existing infrastructure provided by a former railway track, running alongside an elevated section of the London Overground. Just a few miles up the road is the beautiful Parkland Walk, connecting Finsbury Park to Highgate, which has similarly taken a section of old railway line and transformed it into a nature reserve, complete with erstwhile platforms and garishly decorated tunnels, spray-painted by local graffiti artists. But the Camden Highline will differ by virtue of its height – floating high above street level. It draws inspiration from the original New York High Line, another disused railway track-turned-green corridor running along the west side of Manhattan, which was itself inspired by Paris's *Promenade plantée* (tree-lined walkway), completed in 1993.

While in New York for work in October 2017 I visited the High Line for myself. Its structure combines pebble-dash boardwalks and sections of old metal rails with a vast array of plants that bleed through the gaps and add softness to the stark and overbearing landscape. A mixture of grasses and wildflowers, bushes, shrubs and small shady birch trees crowd along its length, the looming warehouses and gleaming skyscrapers on either side creating the impression of an almighty urban canyon. In New York, a city of almost incomprehensible scale, people will now pay millions of dollars to live above, and gaze down on the High Line, a ribbon of green replacing the trains that once hurtled to and fro. This is an often-unintended consequence resulting from the creation of new urban green spaces – the opening of the High Line in 2009 quickly sent adjacent property values spiralling, pricing out many of the local people it was designed to serve.

While too lofty to attract New York's larger urban mammals, such as grey squirrels, raccoons and foxes, the High Line now provides a valuable habitat for pollinators. The plants – a metropolitan melange of native and non-native species – have been curated to provide blooms throughout the growing season, and researchers from the American Museum of Natural History have already recorded thirty-three different wild bee species in total. The horticulture team have modified their annual spring cut-back activities in response to these findings, leaving old stems standing to allow nesting bees to emerge, as well as adding 'bee hotels', made from hollow canes and plant stems to provide additional nesting sites.

'We have incredible biodiversity in terms of the range of species up there, and yet it's also a beautiful garden,' James Corner tells me. James is the Founder of James Corner Field Operations, the landscape architecture firm that was responsible for conceiving the New York High Line and has also been appointed to design the Camden Highline. James grew up near Manchester, but he moved to New York over thirty-five years ago and has lived there ever since. 'It's an extraordinarily rich habitat,' he continues, 'with perennials, shrubs, trees, woodland thickets that are managed in a very particular way so they offer surprise and beauty, but at the same time, they can support wildlife.'

Prior to the creation of the High Line, the New York Parks Department worked from an extremely limited planting palette of around ten trees and shrubs to reduce the need for maintenance. But a vast array of resilient species that can thrive in an urban environment were selected for the High Line; 'We have something like 4,000 species of bulbs, grasses, flowers, shrubs, trees, fungi,' James explains, 'but because it's a dynamic system it all works together and looks after itself.'

That being said, James does admit that the High Line has fifteen full-time gardeners, whose job it is to prevent a single species from overwhelming the available space and to keep the paths clear. There will always be trade-offs in what is ultimately a human amenity first, wildlife habitat second, and this remains the case for most of our urban green spaces. 'I think it's a bit naive to think that it can just all go wild,' he adds; 'it's not perhaps as authentically wild as it might be, but it's an urban kind of wild space.'

Meanwhile, my home turf in Camden feels a million miles away from the mighty urban metropolis of Manhattan, and yet here too a Highline could offer much to both wildlife and people, including a different perspective and a new way of navigating the city. 'It's like being in a helicopter,' Simon claims, 'you barely have any time to clock where you are before you're at the next place. When you go as the crow flies, it's much quicker to get around!'

Simon and his team have already raised enough money to cover the design and planning approval process for the Highline. The next phase will be to fundraise a further £40 million to build it, 'which is less than it cost to not build the Garden Bridge', Simon jokes, referring to the ill-fated vanity project backed by Boris Johnson during his time as London Mayor on which £53 million was spent, including £43 million of public money, without a single brick being laid.

Unlike the Garden Bridge, the Camden Highline will benefit local people, especially those who have little access to nature at present, Simon argues. The proposed route runs through four densely populated housing estates, including Maiden Lane, a notorious no-go-zone in the 1980s and early 1990s, with its post-war-blocky low-rise flats and houses. With the opening of the Highline, 10,000 additional people will meet the current Mayor's aspiration for all Londoners to live within 400 metres of a green space.

We follow the route of the track above as closely as we can from ground level. Spilling over its rusting blue sides are brambles, sycamore, ash and large quantities of buddleia. Buddleia is a prolific plant, native to Japan and the Chinese provinces of Sichuan and Hubei. It has long been popular with gardeners for its vibrant colours – white, purple, blue, pink, yellow, orange, red. Lauded for attracting butterflies during its flowering period, garden varieties have names such as 'sungold', 'attraction',' pink delight' and 'summer beauty'. But its highly dispersible seeds have led to it escaping from gardens and taking over waste grounds and railways, where it outcompetes and engulfs our native vegetation, reducing biodiversity as a result.

Mathew, ever pragmatic, observes that although it would be possible to leave the Highline entirely to nature, allowing it to exclusively self-seed, this would probably result in a similar picture to what we can see right now – a dominance of buddleia and brambles. 'I love it, it's totally wild, but I expect most people don't want that,' he recognises, 'so a mixture of planting, ornament and natural colonisation is probably best.' It is also, he adds, a good idea to think about which species are known to thrive in London, and to adjust the approach based on what succeeds over time – nature is dynamic, after all, perhaps even more so in urban environments.

Of course, planting isn't the only tool at our disposal in a space like this. We discuss the relative merits of adding dead hedges, bug hotels and log piles, or even rain-fed shallow pools, providing watering holes for birds, bats and invertebrates. Bird boxes and bat boxes could be mounted on the Highline's steep sides. Sometimes it's also about what we don't do, as well as the actions we take – given its proximity to the neighbouring blue corridor of the Regent's Canal, I suggest that consideration should be given to the insects, birds and bats that might use the Highline to commute at night by keeping any lighting to a minimum. This shouldn't be too much of an issue, Simon replies, seeing as the Highline will only operate during park hours, from dawn till dusk, and bright lights would interfere with the trains on the parallel track.

If the Camden Highline succeeds in becoming a living, breathing reality it will be down to the hard work of local people, coming together for the benefit of both wildlife and their community. Through our experiences of the coronavirus pandemic, we've finally awakened to just how important our green spaces are. As people flooded our cities' parks, gardens and towpaths, more grateful than ever for the respite they afforded, it also highlighted existing inequities that had long been bubbling beneath the surface. Initiatives like the Camden Highline aren't a 'nice to have' – they're a 'need to have'.

Having completed the route, Mathew and I part ways with Simon at the old Maiden Lane Station on York Way, and head together towards King's Cross. We discuss what the next chapter may have in store for urban wildlife. What might the cities of the future look like, and how could they incorporate more spaces for nature within their design?

We imagine a world where every available roof is an island of green, like those pioneered by Dusty, buzzing with insects through spring and summer, cleansing the air and reducing flooding by trapping rainwater during heavy downpours. Where every wall is an opportunity for vertical planting, as advocated by Dan Raven-Ellison, who envisions our cities as national parks.

Where our buildings are constructed both soundly and efficiently, with inbuilt technology to improve the quality of our water, soil and air. Where intelligent systems pool our resources to vastly reduce the number of vehicles moving goods and people, and electric autonomous transport removes the need for individual car ownership.

Where vehicles no longer sit idle on concrete and tarmac, and car parks and driveways are given over to nature. Where our redundant roads are repurposed as linear parks, green threads that run through the city – a network of high lines and low lines, creating new habitats up above for raptors, ravens, swifts, pipistrelles and noctules, and connecting those isolated populations down below of ground-dwelling creatures like water voles, hedgehogs, slow-worms and newts.

Where sustainable drainage solutions combine with blue

spaces, like those being trialled in Glasgow, to restore our wetlands and waterways. Where streetlights adjust and brighten according to need so as not to crowd out other species with light pollution. Where electrification reduces the constant roar of traffic to the faintest purr, reconnecting us with the sounds of nature and allowing other wild creatures to communicate with one another once again.

It's also vital that we protect those remaining historic and industrial landscapes that are of such importance to certain specialised species – the former brickworks in Peterborough that's home to the largest population of great crested newts in Europe, the clifftops just outside Edinburgh city centre that provide nesting sites for raptors and ravens, the woodlands filled with red squirrels in Formby and the acid grasslands of south London's Richmond Park.

In these wild cities of the future, everyone will have a stake in the success of their local wildlife and this connection will bring with it an awareness and a desire to protect it for future generations.

The collective voices of large-scale organisations such as the Wildlife Trusts, the National Trust and the RSPB, along with those of smaller charities like Froglife and Buglife and movements like Extinction Rebellion, do much to raise awareness and spread the message that we need to take urban nature seriously. But they can't achieve their goals without the help of people on the ground taking action to protect our existing biodiversity and building new habitats in their neighbourhoods.

People like Alison, a constant thorn in her local council's side every time planning officers try to encroach on remaining fragments of habitat. Or Karen, who sparks social media fury every time she hears of a fox cull, while continuing to surreptitiously feed her local pigeons. Or George, climbing up and abseiling down the cliffs of south-east Scotland to ring the chicks of peregrines and ravens. Or Morgan and Scott, kayaking through caves in the snow to monitor their local bat populations. Or Sophie, on her one-woman mission to protect the lowly weed. Or Craig, without whom no one would ever have known that his local stretch of river was being polluted. Or Bobby in New York, working against all odds to transform the public image of the lowly sewer rat. It's the ecologists, the park wardens, the academics, the local wildlife enthusiasts – many of these people spend their day jobs battling to protect our wildlife, and then they continue the fight in their free time too.

But we shouldn't leave it all up to them. We've been transforming the landscapes around us from the day we set foot on this planet and it's about time we did so mindfully, with the ultimate goal of making our cities greener through every possible means. The solutions for embedding nature into the urban environment already exist. It's now about finding the will – political, cultural, societal – to prioritise the needs of nature, of which we too are a part. Urban wildlife is counting on us. The question is what we do next.

A Manifesto for Urban Wildlife

1.   Raise awareness – spread the word about nature in our cities

2.   Make it green – boost vegetation using wildlife-friendly planting, reverse the trend of greying gardens by removing AstroTurf, decking and paving

3.   Just add water – create new blue spaces, big and small

4.   Let it grow – relax mowing regimes, leave self-seeded plants to bloom

5.   Reuse and recycle – use old branches to build dead hedges and log piles, create compost heaps and tub ponds

6.   Be responsible with pets – don't leave unleashed dogs unattended and bring cats in at night

7.   Create connections – use green and blue corridors to connect habitats

8.   Protect what we have – conserve remaining historic spaces that support specialised species

9.   Ditch the weedkiller – avoid using harmful herbicides and pesticides

10. Build better – seal unused pipes and secure our buildings to keep out unwanted visitors

11. Plant on different planes – unlock the potential of underused spaces with green roofs and vertical planting

12. Reduce light pollution – switch off if you can, avoid harsh-coloured bulbs and adopt warmer, gentler lighting at night

13. Give wildlife a home – incorporate niches for wildlife into our structures with swift bricks and bird and bat boxes

14. Campaign for nature – hold local councils, big business and government accountable for the future of our urban wildlife

15. Live and let live – celebrate whatever succeeds in the city

Acknowledgements

This book wouldn't exist without a collection of extraordinary people who, on top of dedicating much of their lives to securing the future of urban nature, took the time to share their wild cities with me. Rather than viewing the natural world in isolation – reserved for those able to escape urbanity in search of a mythical wilderness – they are focused on opening nature up and improving access for others.

These people include Mathew Frith of the London Wildlife Trust, who has done more than most to promote our capital's urban wildlife and who gave up a large chunk of his time to cast his knowledgeable eye over my manuscript, for which I am hugely grateful. Then there's John Tweddle at the Natural History Museum, Daniel Raven-Ellison with his National Park City campaign, Sophie Leguil of 'More than Weeds', Tom Bolton, chronicler of London's lost rivers and David Lindo, a.k.a. the Urban Birder.

There's Karen Heath, friend and saviour to foxes, cats and pigeons, Helen Shore and Mike Priaulx, champions of London's high-flying swifts, Craig Macadam of Buglife, Emily Millhouse of Froglife and National Trust Red Squirrel Officer Rachel Cripps. There's the unremitting bat protectors – Alison Fure in London and Morgan Hughes and Scott Brown of the Birmingham and Black Country Bat Group – and the indefatigable raptor defenders – Stuart Harrington of the London Peregrine Partnership and George Smith of the Scottish Raptor Study Group.

There are the academics and scientists, too, furthering our knowledge and understanding of urban wildlife: Dr Nigel Reeve, tracking the hedgehogs of Regent's Park, Kat Fingland, studying urban red squirrels, Professor Timothy Roper and Dr Bryony Tolhurst, mammalian experts and badger connoisseurs. There's Dr Dom McCafferty, who has joined forces with Cath Scott at Glasgow City Council to learn more about Glasgow's (non-)water voles, Dr Kate Byrne, who first discovered the London Underground mosquito, and Dr Bobby Corrigan, the world's foremost urban rodentologist.

There are the custodians of our urban green spaces and the wildlife that lives within them – Tony Hatton, Mark Rowe and Bryony Cross of the Royal Parks, Ross Edgar of Hampton Nature Reserve, Paul Wilkinson of the Canal and River Trust, and Christopher Skaife (and his ravens) at the Tower of London.

And there are those who imagine, design and create new spaces for urban nature: Simon Pitkeathley and James Corner of the Camden Highline, Camilo Brokaw at Community Rewilding in Glasgow, Gideon Corby and Esther Adelman, a.k.a. the Wildlife Gardeners of Haggerston, and of course Dusty Gedge with his infectious enthusiasm for green roofs.

311

There are also many more who supported my research for this book, including Dr Jane Sidell of Historic England, who helped me to imagine what a prehistoric London might have looked like, Davy Brown of RatDetection.com, who took me out ratting in Buckinghamshire, and Menno Schilthuizen, whose book, *Darwin Comes to Town*, was a huge source of inspiration. I'd also like to thank Professor David Reznick at the University of California Riverside, Dr Ruth Rivkin at the University of Toronto, wildlife photographer and filmmaker Sam Rowley, Heidi Alexander, London's Deputy Mayor for Transport, Shirley Rodrigues, London's Deputy Mayor for Environment and Energy, Graham Tindal, Senior Arboriculture and Landscape Specialist at TfL, Florin Feneru at the Angela Marmont Centre for UK Biodiversity, Zoe Smith at the Hawk and Owl Trust, Mandy Rudd and Maria Longley at GiGL, guerrilla gardener John Welsh, archaeologist Jon Cotton, ecological surveyor Tim Hextell, Dr Luke Dixon at the Bee Friendly Trust and DC Sarah Bailey of the Metropolitan Police's Wildlife Crime Unit.

Equally, this book would never have come into being were it not for my agent, Lucy Morris, who dropped me an email out of the blue asking if I ever fancied 'talking books', and then went on to give me the guidance, self-belief and support to write my own. Or my editor Pippa Wright, who preempted *Wild City* after receiving my proposal during the first lockdown, and who has championed it ever since. From start to finish, Pippa has always understood and shared my vision for this book.

312

ACKNOWLEDGEMENTS

I must also thank my copyeditor Linden Lawson for her eagle-eyed edits and her appreciation of semicolons and em-dashes, as well as Francesca Pearce and Brittany Sankey, who, as I write, are working on an impressive publicity and marketing plan, Sarah Fortune and Rosie Pearce in the project editorial team and Katie Horrocks in production. I'm hugely grateful to Andrew Davis, too, for the beautiful cover and internal illustrations he created for this book, along with Rachael Lancaster, Orion's in-house designer, and text designer Helen Ewing.

Finally, my friends and family (between whom I make little distinction) have been a huge support throughout. None more so than my father, Christopher Reid, who has been my constant sounding board, reader, fellow urban nature enthusiast and companion on many of my trips; who helped me learn how to write in the first place, while introducing me to such wonders as the baby tawny owls in our local woods and the common lizards that basked on railway sleepers.

We lost my mother, Sara Wilkinson, a full decade ago, but it was she who inspired and cultivated my love of the natural world and the creatures that inhabit it. She foraged for toadstools, collected bird skulls, encased small plants and insects in resin and filled my childhood home with a colourful menagerie of dogs, cats, mice, rats, hamsters, ducks, giant rabbits, guinea pigs, tropical fish and a green-cheeked conure named Mrs Weeks.

Then there's my childhood friend and confidant Ceri Law, who has always been by my side and was an early reader of this book, along with my stepmother Liz Miles, whose love and encouragement has been a constant source of strength. And my dear friends Emily Hawkerr, Kate O'Hagan, Becca Warner, Frankie Jackson and Kuki Thanki, who were kind enough to read through and sense-check individual chapters for me, Tom Pullen, who took my publicity shots, and all of my neighbours at the Red Lion, who have been a vital source of connection and friendship throughout the pandemic and beyond.

Last, but certainly not least, Ben has been an unwavering partner, companion and advocate, stepping well outside his comfort zone to sit for an hour and a half in cross-legged silence outside a badger sett, scouring the bushes for an animal that failed to emerge. Or waiting, patiently for the most part, while I take yet another picture of a local pigeon. And all the while continuing to tell anyone and everyone about this book and why they should read it.

Further Reading

Bolton, Tom, *London's Lost Rivers: A Walker's Guide, Volume One* (London: Strange Attractor Press, 2011)

Foster, Charles, *Being A Beast: An intimate and radical look at nature* (London: Profile Books, 2016)

Goode, David, *Nature in Towns and Cities*, Collins New Naturalist Library, Book 127 (London: Harper Collins, 2014)

Goulson, Dave, *A Sting in the Tale: My Adventures with Bumblebees* (London: Vintage, 2014)

Jenkins, Alan, *Wildlife in the City; Animals, birds, reptiles, insects and plants in an urban landscape* (Exeter: Webb & Bower, 1982)

Johnson, Nathanael, *Unseen City: The majesty of pigeons, the discreet charm of snails & other wonders of the urban wilderness* (New York: Rodale, 2016)

Klein, Naomi, *This Changes Everything: Capitalism vs. the Climate* (London: Penguin, 2015)

Kolbert, Elizabeth, *The Sixth Extinction: An Unnatural History* (London: Bloomsbury, 2014)

Lindo, David, *How to Be an Urban Birder* (Princeton, NJ: Princeton University Press, 2018)

Mabey, Richard, *The Unofficial Countryside* (Stanbridge, Dorset: Little Toller Books, 2010)

Mabey, Richard, *Weeds* (London: Profile Books, 2010)

McDonnell, Ian and Adès, Harry, *A Field Guide to East London Wildlife* (London: Hoxton Mini Press, 2014)

Moss, Stephen, *The Accidental Countryside: Hidden Havens for Britain's Wildlife* (London: Guardian Faber, 2020)

Nicholls, Henry, *The Galápagos: A Natural History* (London: Profile Books, 2014)

Orwell, George, *Some Thoughts on the Common Toad* (London: Penguin Classics, 2010)

Packham, Chris, *Chris Packham's Wild Side of Town* (London: Bloomsbury, 2015)

Roper, Timothy J., *Badger* (London: Collins New Naturalist Library, 2010)

Ross, Cathy and Clark, John, *London: The Illustrated History* (London: Penguin Books, 2011)

Schilthuizen, Menno, *Darwin Comes to Town: How the Urban Jungle Drives Evolution* (London: Quercus Editions, 2018)

Skaife, Christopher, *The Ravenmaster* (London: HarperCollins, 2018)

Sullivan, Robert, *Rats: A Year with New York's Most Unwanted Inhabitants* (London: Granta Books, 2005)

Weightman, Gavin & Birkhead, Mike, *City Safari; Wildlife in London* (London: Sidgwick & Jackson, 1986)

Woolfson, Esther, *Corvus: A Life With Birds* (London: Granta Books, 2008)

Woolfson, Esther, *Field Notes From a Hidden City: An Urban Nature Diary* (London: Granta Books, 2013)